NO SHADOWS IN THE DESERT

NO SHADOWS
IN THE
DESERT

MURDER, VENGEANCE,
AND ESPIONAGE
IN THE WAR
AGAINST ISIS

SAMUEL M. KATZ

HANOVER
SQUARE
PRESS

**HANOVER
SQUARE
PRESS™**

ISBN-13: 978-1-335-01383-5

No Shadows in the Desert: Murder, Vengeance, and Espionage in the War Against ISIS

This edition published by arrangement with Harlequin Books S.A.

Library of Congress Cataloging-in-Publication Data has been applied for.

Hanover Square Press
22 Adelaide St. West, 40th Floor
Toronto, Ontario M5H 4E3, Canada
HanoverSqPress.com
BookClubbish.com

Printed in U.S.A.

This book is dedicated to the memory of Yves Debay—
a friend, colleague, soldier, adventurer and amazing combat
photographer—killed by a sniper's bullet in Syria

NO SHADOWS
IN THE
DESERT

TABLE OF CONTENTS

GLOSSARY

1ˢᵗ SFOD-Delta: US Army 1st Special Forces Operational Detachment-Delta, "Delta Force" counterterrorist unit

22 Special Air Service: British Army 22nd Special Air Service Regiment

AQI: al-Qaeda in Iraq

ASL: above sea level

CANSOFCOM: Canadian Special Operations Command

CENTCOM: US Department of Defense Central Command

CIA: Central Intelligence Agency

CJTF-OIR: Combined Joint Task Force—Operation Inherent Resolve

CAOC: Combined Air Operations Center (al-Udeid Air Base, Qatar)

CSAR: combat search and rescue

CTB-71: Jordan's counterterrorist battalion

CTD: the General Intelligence Directorate Counterterrorism Division

Da'esh: the colloquial and derogatory Arabic name for ISIL/ISIS (the Islamic State)

DEVGRU: Naval Special Warfare Development Group, SEAL Team SIX counterterrorist unit

DIA: Defense Intelligence Agency

DGSE: the Directorate-General for External Security, France's international espionage service

ELINT: electronic intelligence

FSA: Free Syrian Army

GID: General Intelligence Department

GIGN: Groupe d'intervention de la Gendarmerie nationale, France's counterterrorism unit

HMMWV: High Mobility Multipurpose Wheeled Vehicle, a Humvee

HUMINT: human intelligence

IED: improvised explosive device

ICRC: International Red Cross and Crescent Movement

ISIL/ISIS: Islamic State in Iraq and the Levant / Islamic State in Iraq and Syria

JAF: Jordanian Armed Forces

JSOC: US Joint Special Operations Command

JTF-2: Canada's Joint Task Force 2 counterterrorist unit

MI6: Military Intelligence Section 6, the UK's foreign intelligence service; also known as SIS, or the Secret Intelligence Service

MIT: Millî İstihbarat Teşkilatı, the Turkish national intelligence service

MRAP: Mine-Resistant Ambush-Protected armored vehicles

NSA: National Security Agency

PHOTINT: photo intelligence

RJAF: Royal Jordanian Air Force

RJSOCOM: Royal Jordanian Special Operations Command

ROEs: rules of engagement

SCIF: Sensitive Compartmented Information Facility

SDF: Syrian Democratic Forces

SIGINT: signal intelligence

TDY: temporary duty assignment

TTIC: CIA Terrorist Threat Integration Center

UNHCR: United Nations High Commissioner for Refugees

USSOCOM: United States Special Operations Command

XO: executive officer, second-in-command

YPG: People's Protection Units, the Syrian Kurdish militia making up much of the Syrian Democratic Forces pro-Western militia

OCTOBER 26, 2019

The night of October 26–27, 2019—
Barisha, Idlib Governorate

ABU BAKR AL-BAGHDADI and his family slept soundly in their beds, protected by several handpicked gunmen who patrolled the grounds of the nondescript compound near the Turkish frontier in northern Syria where the self-proclaimed, so-called Caliph was hiding. The bodyguards did not have to be told that danger lurked in the darkness—the "Ninjas," the local slang for the enemy's special operations forces, preferred to launch their daring raids under the cover of a night sky. The sentries moved about cautiously—avoiding areas near the main gate and the house that were booby-trapped with explosives—and they listened attentively to the winds that howled across the hilltops in order to dissect anything was out of the ordinary. They heard nothing, not even the forewarning of a mountain dog barking at the shadows. The night was silent, and the ISIS gunmen were determined to make it safely through to dawn's first light and morning prayers.

But close to 100 operators from the U.S. Army's Combat

Applications Group, one of the mysterious cover names for the 1st Special Forces Operational Detachment-Delta, along with elements of the 75th Ranger Regiment (Airborne) had already converged on the lair, concealed in stealth and the daring black of night. They had been flown to Barisha from a staging area in Iraq, near Erbil, inside the Kurdistan Regional Government, courtesy of eight helicopters—Chinooks and Black Hawks— from the elite 160th Special Operations Aviation Regiment (Airborne) known as the Night Stalkers, and their seventy- minute-long path in the darkness had been illuminated by the best type and most reliable form of actionable intelligence— the human kind. A source entrenched inside al-Baghdadi's entourage had been turned and he provided accurate and up- to-date information on Baghdadi's whereabouts and movements. The raid had been in the works for close to two weeks.

The Delta Strike Force landed near the compound and quickly surrounded the perimeter. An Arabic-speaker used a megaphone to plead with those behind the walls to come out and surrender, but gunfire erupted. Delta breachers, having been told that the main iron gate was wired with an IED, blew a hole in a spot along the fence in order to gain access to al-Baghdadi's compound. The rest was merely a matter of pinpoint precision and the kind of tactical proficiency few in this world possess, plus raw and undefinable courage. While one Delta team disposed of al-Baghdadi's sentries with a flurry of fire, an element of the raiding party seized two high-ranking ISIS prisoners; another team of Delta operators also whisked eleven children to safety and out of the kill zone. Baghdadi, awakened by the blasts and cadence of selective automatic fire, raced toward a ditch that led to an irrigation tunnel with two children in tow. Delta operators and K-9 handlers followed in close pursuit. Cornered underground, al-Baghdadi reportedly detonated a suicide vest he wore strapped to his torso to ensure that he wouldn't be taken alive, killing himself and the two children.

Abu Bakr al-Baghdadi was the world's most wanted terrorist, a self-anointed king who inspired countless suicidal subjects. But by the time he detonated the capture-proof explosive payload, his roundtable of lieutenants, the men who built the so-called Caliphate and led his army that shocked the world with their unimaginable crimes, had been killed in a combined Jordanian and American intelligence and special operations campaign to avenge the life of a young F-16 pilot whose ghastly murder was a turning point in the global war on terror.

This is the remarkable story of how the so-called Islamic State in Iraq and Syria came to an end.

AUTHOR NOTE

It is mine to avenge; I will repay. In due time their foot will slip;
their day of disaster is near and their doom rushes upon them.

—*Deuteronomy* 32:35

THIS BOOK IS about vengeance—the no-holds-barred, eye-for-an-eye variety of retribution that only nation-states can administer. The capture, torture and immolation of a young Jordanian F-16 pilot by the so-called Islamic State*, a crime that was filmed and then tweeted to the world, set in motion a top secret vendetta carried out by the spies and covert operatives from the Hashemite Kingdom of Jordan, the United States and a global coalition of nations that was meant to punish and obliterate men capable of such evil—from the low-level thug who beat the airman with his rifle, to the men in charge who orchestrated such a display of barbarity. The elimination of the Islamic State's top leadership proved to be a turning point that marked the beginning of the end of the Caliphate.

---

* In the Arab and Islamic world, the term "Islamic State" is considered an insult to the religion and its followers. It is used in the pages of this books to avoid confusion between the terms ISIS and ISIL that are frequently used to describe the terrorist group that most of the Middle East simply refers to as Da'esh.

The changing of fortunes on the battlefield—the coalition's recapture of territory and the liberation of hundreds of thousands of grateful people from the Islamic State's grip—and the ultimate eradication of the physical Caliphate has not eradicated the twisted interpretation of Islam that led men and women from the four corners of the world to travel to Iraq and Syria to become foot soldiers marching toward the apocalypse in the first place. That mind-set still exists; it flourishes, in fact. What happened inside the caves of Afghanistan with al-Qaeda and then in the killing fields of Iraq and Syria were symptoms of a thousand-year sectarian war that has raged. The battle lines have placed Sunnis against Shiites, fundamentalists against moderates, and fanatics against all that's Western. The fight has turned the religion inside out, and the conflict does not stop just because the shooting has. The military campaign against the Islamic State is over, but the next phase of the war, for the heart and soul of the virtual Caliphate, is about to begin. Many in the intelligence community fear that this new rendition of the fight will be bloodier than ever before. The fight against the khawarij, or outlaws of Islam, as the zealots are known, is global and requires the combined efforts of strategic and dedicated alliances.

This book is also about such an alliance—the military and intelligence bonds between the Hashemite Kingdom of Jordan and the United States. The bilateral ties have lasted for over seventy years; they have transcended politics and presidents. The links between both nations have grown exponentially over the years—especially in the aftermath of the September 11, 2001, attacks when Jordan became a frontline combatant in the global war against terror. The intelligence services and military arms of both nations became inseparable combat partners in the coalition to obliterate the Islamic State.

This book was written with the cooperation of soldiers, intelligence officers and airmen who fought the Islamic State. Many of those interviewed for the book are still in the fight, operating inside Iraq and deep inside the fissured battlefield of Syria. These men and women remain exposed to supreme and permanent risk; they will probably look over their shoulders for the rest of their lives. I have, therefore, taken great care to protect their identities and any element of their tradecraft that could endanger them. I have used pseudonyms when warranted, altered a person's physical description where needed and even changed the location of certain sit-downs and clandestine rendezvous.

This manuscript was also vetted solely for the measure of preserving the operational security of the men and nations involved in this long and brutal war. All opinions, however, are my own.

Samuel M. Katz
New York City, September 2019

PROLOGUE

Slider One-One

THE BLUISH-WHITE MIST of dawn blanketed the tarmac that December morning as it did every day when the sun slowly climbed over the desert hills of northern Jordan. A light eastern breeze sent a chill across the airfield; the whip of the wind howled across the vast open space. It was quite common to see the odd caracal or two scampering across the dunes and embankments to escape the first rays of light that were fast approaching, still hoping to catch a bird or a rat before retreating into the barren habitat of the desert. There were patches of green here and there; even the desert blooms with a blanket of foliage from the rains that begin at the very end of autumn. The emergence of daylight was a magical sight in the desert, even at the remote Muwaffaq Salti Air Base, located just seven miles from the Syrian border, where the roar of jet engines could be heard long before the sun appeared.

Few appreciated the majesty of morning breaking across the sandy hills more than twenty-six-year-old First Lieutenant Moaz al-Kasasbeh,[*] an F-16 pilot in the Royal Jordanian Air Force

---

[*] There have been numerous spelling variations of both the first and last names. For the sake of uniformity and clarity, I have selected the spelling used by the *New York Times.*

No. 1 Fighter Squadron, who gazed at the desert splendor as he sat in his fully armed aircraft awaiting permission to take off. It was 6:30 on the morning of December 24, 2014. He was about to lead a strike against a terrorist weapons depot south of Raqqa, in Syria. It would be his sixth bombing sortie in the two months since the international coalition known as Inherent Resolve had declared war against the Islamic State.

Seventy-two hours had passed since the squadron received its order from the coalition's command center and the strike had been assigned to Moaz and his number two. The planning sessions with the US Air Force and combat units from other nations included intelligence briefings and an analysis of the weather over the target area, which took up much of the day. The preparation also included a secure video link with a US Marine Corps contingent that was responsible for combat search and rescue in case something went very wrong. It had been forty-one years since the Royal Jordanian Air Force, or RJAF, had really been at war, fighting Israel atop the Golan Heights; there had been a few operational missions since—in Iraq and Libya—but they were primarily covert and small in scale. Now the RJAF was at the forefront of a multinational coalition to defeat one of the most lethal terrorist armies ever to emerge, and Moaz felt like a kid once again, barely able to contain his excitement about leading an air strike. "Moaz knew that he was in the process of making history," a squadron senior officer remembered, "and that excited him."[1]

There was little time to sleep the night before the mission— too many things to take care of and too much adrenaline flowing through pilots' bodies. At 5:00 Moaz attended to his morning prayers and then sent a text message to his new bride, Anwar. Faith guided the young pilot, but so did the love he shared for his wife of four months. Moaz looked at Anwar's photos on his smartphone and then tucked it into a side pocket on his khaki

flight suit. Moaz was hoping to message his beloved right before takeoff and thought he'd have time to return it to his locker
afterward.

The No. 1 Squadron commander and his XO walked Moaz
and his flight partner to the briefing room. Some of the pilots
drank small cups of Bedouin coffee laced with cardamom. The
dark, muddy brew was pure rocket fuel and the perfect elixir before a combat mission. Other pilots smoked, creating a stagnant
and suffocating white cloud that hovered stubbornly throughout
the room. Moaz thought that smoking was a filthy habit and he
wasn't a coffee drinker. He welcomed the morning with a glass
of tea with mint leaves. He was very much a Boy Scout when
it came to adult vices.

The squadron briefing room was large enough for six rows of
chairs arranged like a classroom. It was equipped with a projec-
tor and a screen. A wooden podium decorated with the squadron emblem was positioned in the front of the room. A huge
banner bearing the hunter dog image and inscribed with the
words "It was ever incumbent upon Us to give victory to the
believers" from the Koran hung from a wall. The No. 1 Squadron's emblem was a black hunting dog that paid homage to the
British-made Hawker Hunter fighter that served the squadron
during the 1967 Six-Day War; the air base, in fact, was named
after a Hawker Hunter pilot killed in a dogfight with the Israeli
Air Force in 1966. Framed portraits of King Abdullah and King
Hussein were displayed prominently in the war room.

The preflight briefing belonged to the squadron intelligence
and operations officers. Their job was to review the detailed
step-by-step checklist of mission details and to make sure that
the pilots were completely fluent with all the landmarks of their
strike. The briefing included any last-minute details that the pilots needed to be concerned with, such as weather updates, and
last-minute sensitive intelligence on the target and the threats
they could encounter. Part of the briefing was teleconferenced,

with coalition personnel detailing any last-minute changes to the US Marine Corps special ops team that would serve as the combat search and rescue force.

The pilots retrieved their G-suits and gray flight helmets from the ready room, along with their gear bags. Suiting up took time. The pilots had to squeeze into Nomex G-suits, flight boots, harnesses, life preserver units, Nomex gloves and survival vests—complete with an emergency radio transmitter and GPS, commando dagger, and a Glock 19 semiautomatic 9mm pistol. Flying a Mach 2 fighter-bomber like the F-16 might have been high speed and low drag, as a squadron officer commented, but the survival gear that a pilot carried into battle consisted of a sharpened blade and a sidearm.

Moaz's squadron call sign was "Safi." It was his father's name. But the coalition code name for the morning's strike was "Slider." As the flight leader, Moaz's radio code name was Slider One-One; the second aircraft, piloted by Lieutenant Salim,[*] was code-named Slider One-Two.

Moaz and Salim were driven to the revetments at a remote stretch of the air base where the two F-16As were readied. There was silence on the white minibus during the five-minute drive— both pilots were already in mission focus, and they looked out the window deep in thought. The pilots then performed a walk-around inspection, checking the ordnance and the fuselage. It was the standard operating procedure. Aircrews had already prepped the F-16s for battle with a ground-attack ordnance configuration that was known as "organic": four MK-82 500-pound bombs were affixed to pylons—two underneath each wing—and two AIM-9 Sidewinder air-to-air missiles were on the wingtips, just in case Syrian Air Force fighters tried to intercept the tandem over the target.

---

[*] The name Salim is a pseudonym. The pilot's real name is concealed for operational security.

The aircraft were camouflaged in a subdued two-tone scheme of gray—a Jordanian flag adorned the tail, and both wings were marked by red, black, green and white RJAF roundels. Both pilots climbed the ladders affixed to the left side of their respective aircraft as the crew chiefs and ground crews scurried about. There isn't much wiggle room for anyone sitting inside the narrow confines of the F-16 cockpit: the ejector seat was designed at a small recline so that the pilot's legs could fit conveniently and comfortably inside. At five feet, eight inches tall, Moaz fit snuggly behind the space-age controls—an inch or two taller and a few pounds more, and it would have been a tight squeeze. He checked his G-suit once again, and his harness. He fastened his harness slightly, and then checked the maps and coordinates held on his legs and visible through transparent slots in his G-suit. Moaz wiggled his fingers into his Nomex gloves, maneuvering forcefully until the fit was to his liking. He then checked his systems, his displays and his controls. Finally, he tested his pedals and his throttle. All systems checked. Moaz lowered his visor and issued the thumbs-up. The ground crew removed the ladder. The aircraft was ready. Both F-16s taxied to weapons areas where the crew chiefs armed the bombs and missiles. It was then a short taxi toward the takeoff line where the aircraft would wait for the green light to launch.

Moaz was slated to fly the following day, on Christmas, but in the hope of enjoying a long weekend with his new wife, the young pilot submitted a formal request to swap places with another pilot. Lieutenant Colonel Ali,* the squadron commander, saw nothing wrong with approving the change in the roster—it was routine. Moaz would be debriefed following the sortie, he'd shower and change, and then work on transportation so that he could be in his new home before dusk. But that was later. On the tarmac Moaz took a deep breath, gazed across the runway

---

* A pseudonym. The commander's real name is concealed for operational security.

and glanced back at Lieutenant Salim, who was positioned behind him on the apron.

"Slider One-One to Tower," Moaz radioed in over the secure
communications network. "Ready for departure." The control
tower, staffed with RJAF and US Air Force personnel, checked
with the coalition frequency to receive the required and final
mission authorization. "Slider One-One, over," the traffic control officer responded, "departure confirmed."

First Lieutenant Moaz al-Kasasbeh fired up his afterburners
and aimed a thumbs-up to the control tower. Seconds later Slider
One-One launched his aircraft straight into the winter sun toward an easterly course. Slider One-Two was close behind. The
plan was to be back at base before lunch.

BOOK ONE

A NEW STORM OVER THE HORIZON

1

What Could Go Wrong?

Real victories are those that protect human life, not those that result from its destruction or emerge from its ashes.

—**King Hussein**[2]

SUSAN RICE, PRESIDENT Barack Obama's national security advisor, was a frequent visitor to the capital cities and sprawling US-built air bases east of the Jordan River and west of the Persian Gulf in 2013 and 2014. Rice always traveled with a platoon of faithful staffers. Her roadshow came with a supporting cast of go-to people: subject-matter experts with highly focused tactical and strategic perspectives who worked for the Central Intelligence Agency and units from the covert end of America's special operation community that was never mentioned in public. Rice's entourage was large—too large, sometimes—for the converted passenger jets from the US Air Force Special Air Mission 89th Airlift Wing, which transported cabinet members and congressional delegations to and from the world's trouble spots; sometimes, conditions necessitated the use of military cargo craft. The aircraft were often small and uncomfortable, especially considering that the American national security advisor and her delegation traveled with a phalanx of heavily armed, earpiece-wearing Secret Service agents.

Rice was one of President Obama's most loyal aides and trusted advisors. She had been by the president's side from his meteoric race to the White House all the way to her appointment as United Nations ambassador. Educated at Stanford and Oxford, Rice was a Rhodes scholar known to have an academic perspective about the world and America's role in it. In fact, when President Obama introduced her as his national security advisor in the sun-soaked splendor of the White House Rose Garden, he made a point of saying, "With her background as a scholar, Susan understands that there's no substitute for American leadership."[3]

When Rice traveled to the Middle East to strategize with her counterparts, protocol warranted that the top-ranking official from the host nation usually sat at the head of the table. But Rice left no doubt to those in attendance that they were guests in her court. The talking points at these high-level meetings were always written according to American concerns and President Obama's foreign policy agenda; there were no doubts that she was speaking with the president's voice. But Rice's tone was known to cut through the pleasantries of standard diplomatic practice. She had a unique know-it-all style about her, a participant at one of these meetings remembered. She lectured, to the dismay of some, when listening would have been better; it made some people who would be in the room with her uncomfortable. She was impatient, others had noted, and she wasn't fond of bullshit or stupidity.[4]

Some of the regional leaders and spymasters that met in this capital or the other were put off by her certainty that she and she alone knew best. It had nothing to do with the fact that Rice was a woman, though. Women had sat at the head of such polished cherrywood tables before, of course—Secretary of State

Hillary Clinton and Gina Haspel,* the acting head of the CIA's Clandestine Service, to name a couple. But once again the region was at war, and Middle Eastern conflicts, especially those involving religion, were historically the business of men.

There was a time when frequent high-level gatherings in Arab capitals that involved the American president's national security advisor would have been dominated by talk of Israel and the Palestinians. Then there was a time when discussions of Iraq and Iran dominated the agenda. But the onset of the Arab Spring had changed the Middle East's talking points forever. The priorities and concerns of the nation-states that were in the American and Western orbit had been sharpened and localized. There had been regime changes in Tunisia and Yemen. The fervor and zeal of Arab Spring protests toppled Egyptian president Hosni Mubarak, leader of the largest of all Arab States and the president of America's anchor of regional stability. In October 2011, with the help of NATO air strikes, Muammar Qaddafi was ousted as Libya's strongman, ending forty-one years of tumultuous rule of the oil-rich North African nation. There were protests and rioting in fourteen Arab countries.

It took the Tunisians four weeks to rid the country of Zine el-Abidine, the country's dictator for twenty-three years.[5] The Syrian people weren't that lucky.

The opening barrage in the Syrian Civil War came from the crashing sounds of bones breaking inside a dungeon in the southern city of Daraa one sunny morning in March 2011. State security agents arrested fifteen young boys who had spray painted the phrase "the people want to topple the regime" on their school wall. Rather than being turned over to their parents, the boys—whose ages ranged from ten to fifteen—were tortured with ghoulish cruelty. According to some reports, they were

---

* At the time of this book's writing, Gina Haspel is the Director of Central Intelligence, the first woman to ever lead the CIA.

savagely beaten; their fingernails were pulled out.[6] News of the boys' arrests spread throughout the town and then throughout the country. Prominent political prisoners staged hunger strikes. Leaders from two Syrian Kurdish factions soon joined in with acts of civil disobedience. President Bashar al-Assad's security forces, known for their heavy hand, overreacted to the large-scale public protests. Scores were arrested, never to be heard from again; many, including women and children, were shot and killed. Eventually the fifteen boys were released, but the spark had ignited decades of rage directed against a brutal regime that represented a minuscule and corrupt minority in a nation that was deeply divided along religious and sectarian lines.

The Syrian Civil War, a byproduct of the Arab Spring, had metastasized into a full-blown global crisis that had the potential to spread to the other nations of the region and embroil Iran, Russia and the United States. President Obama realized that for the bloodshed to end, Syria's dictator, Bashar al-Assad, would have to go. Susan Rice was dispatched on her shuttle strategy sessions to help bring about regime change in Damascus and, at the same time, to keep the turmoil well contained inside Syria's fractured borders. It was a tall order, indeed.

Leaders from all around the world had urged Syria's President Assad to show a delicate hand when dealing with the protesters. The pleas even came from Arab heads of state who realized the combustible potential in a predominantly Sunni land that was ruled by an Alawite* minority allied with Hezbollah and the Islamic Republic of Iran, which also had a rebellious and organized Kurdish separatist population.

Assad declared war on those inside the country who demanded change, escalating the unrest into armed insurrection.

---

* The Alawites are an ethnoreligious group, primarily based in Syria, that have fused multiple theological beliefs and practices together into a highly secretive religious sect that reveres Ali, one of Mohammed's descendants; the Koran is only one of their holy books. The Alawites are fiercely independent and have now allied themselves with Shiite Iran and with Lebanon's Hezbollah.

Rioters were shot and killed; more people disappeared. The response was violent. Secret police installations were overrun and ransacked. Government offices were torched. Scores were settled with brutal disregard for due process or mercy. Political and state security officials were strung up and executed. By the end of the summer, full-scale internecine fighting had erupted throughout the country. The ruling Alawite minority and its security forces came under attack from a hodgepodge of separatist groups representing a myriad of ethnicities, political beliefs and religious zeal in separate parts of the country. Kurdish militias fought in the north; former military men and moderate Sunni forces from the Free Syrian Army fought in the major urban areas of Homs, Aleppo and Damascus. Fundamentalist Sunni forces, including men who fought in Iraq and even Afghanistan with al-Qaeda and with Abu Musab al-Zarqawi's al-Qaeda in Iraq, attacked government forces in southern Syria and along the Iraqi frontier in the no-man's-land of the desert. New groups such as the al-Nusra Front promised to instill a harsh Sunni brand of sharia, or Islamic law, to the lands they liberated and welcomed volunteers and cash. In January 2012, the al-Nusra Front for the Liberation of the Levant made its debut on the Syrian battlefield: on January 6 a suicide car bomb killed twenty-six people in Damascus, and on February 28 there were simultaneous suicide car bombings in Aleppo that killed twenty-eight.[7]

Damascus thought that brute force carried out by the muscle of thousands of Hezbollah volunteers from Lebanon and Revolutionary Guardsmen from Tehran could save the regime's slipping hold over the country. The flow of men, women and children fleeing the fighting soon developed into one of the most horrific human tragedies in recent memory.* Hundreds of thousands of refugees escaped to Turkey in the north and Lebanon to the east. Those who could afford to pay a small fortune booked passage

---

* Sadly, it wasn't until the heartbreaking photograph of the lifeless body of Aylan Kurdi, the Syrian boy who washed up in a Turkish beach in 2015, that the refugee issue attracted temporary global outrage.

on rickety freighters leaving the Mediterranean ports for points in Europe. Most traveled as families, and they hoped that they could make it to Greece or Italy before the ships succumbed to age and rough seas. Men sold their cars and their homes to assemble the necessary cash; women sold their gold jewelry. Smugglers and black marketeers made a fortune.

The majority of those running for their lives headed south to Jordan.

By 2013 there were close to one million Syrian refugees inside Jordan. The refugees fled with only what they could carry on their backs to make it to the holes in the border fence that allowed them to cross into safety—they trekked to the border in the snow-swept subzero winter, and they marched for days under the summer's inferno sun. Survivors of the difficult trek to safety told aid workers that they were sometimes used for target practice by government forces before they reached the frontier. Many spoke of atrocities that had been perpetrated against Sunnis suspected of resisting the government in Damascus; there were executions and reported rapes,[8] and, in some cases, people simply vanished. One of the refugees fleeing the carnage in Syria who found haven in Jordan was Naief Abazid, one of the fifteen teenagers tortured by Assad's secret policemen that precipitated the fissuring of the country.[9]

Jordan was in no position to absorb a million or more refugees a decade after absorbing tens of thousands of refugees who fled the American invasion of Iraq; there were still Palestinian refugee camps in the country, a reminder of the 1948 and 1967 wars. The cost of it all was more than Jordan could bear, and the security threat of so many Syrians who represented a myriad of ideologies and factions was also of critical concern. These political, humanitarian and, most of all, security concerns were of paramount importance to the Obama White House, just as it had been to previous administrations: Jordan was the linchpin of vital American interests in the Middle East.

Jordan was an unlikely US focus. It had no oil, few natural

resources to speak of and little else of monetary value. It did, though, cover an all-important crossroads of barren desert real estate that bordered four influential players in the region: Iraq, Syria, Saudi Arabia and Israel. As far as the State Department saw it, the chances for a comprehensive Middle Eastern peace hinged on Jordanian participation; indeed, the Hashemite Kingdom was one of two Arab states to maintain a peace treaty with the State of Israel. In American eyes—those in the Pentagon and at Foggy Bottom—if Jordan fell, all the other pieces—the American investments and vital American interests in the region—would fall, as well. "If Jordan was safe, the Middle East was stable, and America could rest at ease," said Colonel Khaled,* a retired colonel from the Jordanian special forces and intelligence services. Capitals across the West and inside the Arab world knew that Jordan under the gun was a situation that risked a destructive Middle East–wide conflagration. Jordan was a buffer that helped an unstable region maintain periods of peace and status quo.

Jordan was also a frontline state in the global war on terror. Jordanian intelligence officers and special operations units had served alongside top-tier American and British units in the provinces of Afghanistan and helped to secure key installations.[10] For the Middle East, and America's interests in the region, Jordan's security was a true red line in the sand that the United States was determined never be crossed. The civil war in Syria pushed Jordan—and the region—to the brink, and Washington was determined to avoid collateral damage.

Jordan was already knee-deep in the Syrian conflict. Jordanian intelligence and special operations personnel were working side by side with the Central Intelligence Agency and Britain's MI6 as advisors and language specialists—they spoke the local dialects, understood the nuances of the local customs. The Jordanians understood the tribal mind-set, even inside the opposi-

---

* A pseudonym.

tion groups. "We had intimate comfort with the local landscape," Brigadier General Amir,* a Royal Jordanian Special Operations Command officer, explained. "The Americans and the British didn't have the personnel who truly understood the complex human makeup of who we were dealing with. They didn't know who could be counted on to remain on our side and who we might one day find ourselves fighting against."[11]

Officers and soldiers from the Free Syrian Army, the FSA, were the easiest for the Jordanians to classify as "snow white." The FSA was made of ex-military and former policemen. There were personnel files to review; the Jordanians and others had dossiers on some of the personalities. There was no way to predict what their politics would be when and if the shooting ever stopped and if given the opportunity to govern—whether they would unite the country or, as happened in Iraq and elsewhere, use their power to settle old scores on ethnic and religious lines. But for now the FSA was completely secular, and the covert Jordanian contingent was under strict orders to make sure that weapons would never end up in the hands of Islamic fundamentalists. It was essential, for Jordanian security and stability, that weapons not reach the hands of the extremists.

Not only did Washington and the other NATO nations misread the facts on the ground, but they didn't want the Jordanians—or anyone else—involved. There were concerns in the Western capitals that if additional nations became embroiled in the fighting, the situation could quickly get out of control. The United States and NATO were reticent to commit advisors—and certainly combat units—to the fighting.

Geopolitical concerns muddled efforts to steer the fighting toward the West's advantage. The United States—and some of the NATO powers—insisted that any aid to the Kurds had to be shipped via Baghdad, which was maddening to members of the intelligence and special operations community covertly as-

---

* A pseudonym.

sisting the Kurdish forces and their respective governments. The Baghdad central government had no interest in bolstering its autonomous Kurdish competitors to the north. Iraqi authorities often blocked the shipments of arms and ammunition headed to Erbil—even mobile field hospitals and medical gear—but allowed Iranian military aircraft to land in Baghdad to refuel and then continue to Damascus with weapons and fighters that were battling the Kurds and other rebel forces.[12]

The madness, though, was multidimensional. Other voices urged King Abdullah to become proactive in the civil war and to enlarge the conflict, which, up until this point, was contained inside Syria. John McCain was a frequent visitor to the Hashemite Kingdom and was Jordan's staunchest supporter on Capitol Hill. The Arizona senator was also a good friend to King Abdullah. The veteran Cold War–hawk urged the king to open a second front along the Syrian frontier. The king was not keen to commit the full might of his military and risk a much wider conflict. It was imperative for Jordan that it have a say in what was going to happen.

Instead, the Western powers were adamant that they have the ruling decision in what happened in Syria during the war and once the shooting stopped.

Susan Rice flew to Amman in late 2013 and early 2014 to meet with King Abdullah and the other power brokers in the region. The participants were all met with an honor-guard salute when they exited their armored vehicles in front of the main entrance of al Husseiniya Palace in Amman; the attendees were then ushered to an ornate but secure conference room. Middle Eastern treats and bottles of mineral water were waiting on the table.

King Abdullah welcomed the visiting dignitaries to the roundtable in a set of digital-pattern, desert-camouflage fatigues. The Jordanian commander in chief was the only head of state present with a military pedigree. In fact, before ascending to the throne in 1999, Abdullah hoped to make the military a career. He was a graduate of the British Military Academy at Sandhurst and went on to serve as a tank officer, eventually taking command of the elite 40th Armored Brigade. Abdullah later focused on commando warfare and counterterrorism; he lateralled to the special forces to command a brigade and two years later created a unified special operations command that included CTB-71, Jordan's elite national counterterrorist and hostage-rescue force.

King Abdullah wanted to contain and extinguish the fundamentalist fire before the fanatical terrorism spread to Jordan. Many of the groups, such as the al-Nusra Front, were known fundamentalist Islamic terror groups and had aspirations beyond Syria. The entire situation, designed to precipitate a change of regime in Damascus, had the potential of now spreading from Damascus to the other capitals of the region. Each possible scenario created a logbook of worries.

Much to the surprise of many of the Middle East's leaders, Susan Rice advocated arming all the groups regardless of their ideology and political persuasion.[13] It was an astonishing statement coming from Rice, who represented an administration that always pursued policies of caution. The question of weapons and who'd receive them was a terribly sensitive topic. Some of the men in the room couldn't believe the suggestion, especially since the weapons situation inside Syria had become a free-for-all. In some cases, in fact, militants armed by the Pentagon were fighting those armed by the CIA.[14] There were other intelligence operations, mounted by the Turks, the British, the French and others that were sending arms to specific groups. The map of the modern Middle East was filled with the graveyards of those killed by unrestrained best intentions.

King Abdullah was adamant that not a single bullet, handgun, assault rifle or satchel charge should reach a militant that everyone in the room couldn't say with certainty wasn't a terrorist.

Other Arab leaders and field commanders shared King Abdullah's resolve. There was a sense of pure disconnect in how Washington viewed the tumultuous events transpiring in Syria. What was America's endgame in all of this? they wondered. After all, as Rice had once said, "There's a whole world out there and we've got interests and opportunities in that whole world."[15]

The Amman meeting, designed to build a coordinated consensus among the allies, ended up continuing the status quo of confusion. Syria dominated the discussions but remarkably there wasn't any overall agreement or actionable plan agreed to by all. Dusk had cast a shadow over the Jordanian capital as the various motorcades raced toward hotel suites or the residences at their respective embassies. Some headed toward the military side of the city's Marka Airport, where their official jets waiting.

As one of King Abdullah's political advisors made his way out of the palace, he turned to a colleague in the GID, Jordan's General Intelligence Department, who was also at the meeting, and snickered, "What could go wrong?"

2

The Fall of Mosul

What is postwar Iraq going to look like, with the Kurds and the Sunnis and the Shiites? That's a huge question, to my mind. It really should be part of the overall campaign plan.
  —**General Norman H. Schwarzkopf, talking about the 2003 American-led invasion of Iraq with the *Washington Post*[16]**

LIEUTENANT GENERAL MEHDI Sabah Gharawi, the Iraqi commander of the country's western Nineveh Province, was an imposing figure. He was tall and muscular with a hooking camel-like nose that pressed down on a thick military mustache that covered much of his upper lip. Gharawi walked defiantly and with a devilish grin—the swagger of a powerful man who found perverse pleasure in the use of brute force. He was an incompetent military commander, though, and he was as corrupt as they came. Gharawi's nickname was "General Deftar," named so after Iraq's 10,000-dinar note[17] because he notoriously sold officer commissions to anyone who brought him a suitcase full of cash.

Gharawi learned the art of corruption—and the knuckle-dragging, fingernail-pulling methods to enforce it—while serving as an officer in Saddam Hussein's Republican Guard. Gharawi was Shiite and was one of the very few members of that branch of Islam to serve, let alone climb up the ranks, of the Sunni-dominated Praetorian guard. Iraq's sectarian balance of power, enforced at the end of a bayonet, was turned upside down following the American-led invasion in 2003. The one-time second-class Shiite minority was suddenly propelled into the new ruling elite. Shiite veterans of the Republican Guard with a pedigree of ruthless violence like Gharawi were thought to possess the useful qualities that would be needed to enforce the coalition's pacification of post-Saddam Iraq. Gharawi was recruited into the national police and made a general.

If there was to be a future for Iraq, the country needed men who could unify a multiethnic nation that had for so many years been held together by fear and terror. But old proclivities died hard, and the sectarian violence that many feared came quickly and horrified American officials. Gharawi's Shiite-led security forces were connected to a slew of extrajudicial killings against notable Sunni community and business leaders. Of particular concern was Site 4, an illegal Baghdad jail run by Gharawi, where prisoners, primarily Sunnis, were tortured; some were reportedly sold to local Shiite militias only to never be heard from again.[18] Political officers working out of the US Embassy in Baghdad urged Iraqi prime minister Nouri al-Maliki to arrest Gharawi for the human rights abuses; a special investigative and prosecutorial unit called the Major Crimes Task Force, consisting of both American and Iraqi prosecutors, built a compelling criminal case against Gharawi.[19] He was reported to have personally tortured prisoners.[20]

Maliki refused to sack his controversial, yet reliable, heavy hand. He knew that his trusted Shiite enforcer's talents would be needed later.

In 2011, following the successful surge by US and coalition forces in Iraq to crush a fundamentalist Sunni and pro-Saddam

Baathist uprising, Maliki promoted Gharawi and sent him to run Nineveh Province, 14,100 square miles of oil wealth and religious strife in the western end of the country, abutting most of the Syrian border. "It was," as one former State Department security officer who worked in Baghdad commented, "like sending a pyromaniac to protect a warehouse full of oil-soaked rags."[21]

Of the three and a half million people living in Nineveh Province, two-thirds lived in Mosul, an ancient trading center with thriving economic foundations known for its tolerant, multiethnic composition. "Mosul was," as one local journalist commented, "very much like Aleppo in Syria, or Dubai or, in another context, a Geneva—a city built on banking and trade. The city's markets, banks, infrastructure and opportunities invited the best minds from miles around regardless of their race or religion."[22] Mosul had always been special—a brilliant jewel of fertile lands and protective hills surrounded by a foreboding desert and fractious tribal entities. The city had been a central pillar in the Babylonian and Assyrian Empires and was overrun by countless armies in its long and turbulent history.

Mosul sat on the geographic rift separating the Arabs and the Persians—a seismic fault line that separated Sunni and Shiite. The Ottoman Empire's Sultan Suleiman the Magnificent consolidated his rule over much of the Middle East, including what is now Iraq, in the first of three campaigns to humble the Persian Empire. Mosul became its own province, and Sunnis and Shiites existed peacefully. Mosul, historically, had thrown out the welcome mat for minorities with a penchant for survival and achievement, such as the Kurds, Turks, Yazidis, Armenians and Jews. Situated at a halfway point between the Mediterranean Sea and the Asian trade routes of the Persian Gulf, commerce was at the core of the bustling metropolis, and as long as money changed hands, the city thrived in a multiethnic equilibrium.

Oil and the endless billions of dollars that flowed beneath the ground reinforced Mosul as a major banking and commerce center. The historic enmities between the Sunnis and Shiites were overshadowed by the oil revenues that turned Saddam Hussein's sectarian regime into one flush with cash. A concerted effort by Saddam Hussein to make the region pure "Arab" had forced many of the minorities to flee to the countryside. Still, even during times of peace, Mosul was always a city divided along religious lines. The Sunnis lived on the west bank of the Tigris River, and the Shiites and Kurds lived on the east bank. Mosul was the largest Sunni-majority city in Iraq, but, following the 2003 war, the city boasted the largest Kurdish population living outside the safety of semiautonomous Kurdistan in the north. Offers by the Peshmerga, the principal Kurdish militia, to coordinate operations against the many al-Qaeda-linked cells deeply entrenched inside the major towns and cities of the province were dismissed by Gharawi. Nouri al-Maliki, fearful of Kurdish aspirations, sent Gharawi to Nineveh to keep the Kurds in check.

But the Kurds weren't Baghdad's most urgent problem. A man known to the intelligence dossiers by his nom de guerre of Abu Abdulrahman al-Bilawi was.

Abu Abdulrahman al-Bilawi was a short boulder of a man, made of solid muscle and stubborn resolve. He carried the face of a boxer, one that had been punched in the nose very often. A member of the Albu Bali clan of the Dulaim, the most prominent Sunni tribe in Anbar Province known for producing fierce warriors, al-Bilawi was commissioned from the Iraqi Army's version of West Point in 1993 and eventually reached the rank of captain. He was an infantry battalion commander in 2003 when the Americans invaded.

Al-Bilawi's transformation from a military officer to a terrorist was a direct byproduct of an American attempt in nation building that went horrifically wrong. On May 23, 2003,

Coalition Provisional Authority of Iraq Order Number 2 dissolved the Iraqi armed forces and the organs of state security. That decision was the brainchild of American diplomat L. Paul Bremer III, the administrator of the Coalition Provisional Authority of Iraq, and it was dangerously shortsighted. Overnight over 250,000 soldiers, spies and bureaucrats lost their salaries and pensions. The move was immediate and irreversible. Men who had dedicated their lives to serving Iraq and who were highly trained in the art of war and in counterinsurgency were rendered destitute and futureless, left to worry about how they would feed their families and bitter that their pensions had disappeared with the mere stroke of a pen. Many of those issued their walking papers took up arms with the Baath Party underground and fought the American occupation. Others, like al-Bilawi, helped turn the Sunni fundamentalist movements into formidable combat formations.

Al-Bilawi headed military operations in Abu Musab al-Zarqawi's Tanzim Qaidat al-Jihad fi Bilad al-Rafidayn, the al-Qaeda offshoot in Iraq. Known to the West simply as AQI, this potent and ultimately rogue franchise of Osama bin Laden's Islamist army was a lethal enemy, striking with diabolical consistency and effectiveness and unleashing a murderous string of seemingly unstoppable suicide bombings. AQI was characterized by military discipline and a hard-core cadre of combat-proficient operatives who perpetrated many of the most diabolical strikes launched against American and Iraqi forces.

Coalition forces arrested al-Bilawi on January 27, 2005,[23] as part of the massive American-led effort to locate and terminate Zarqawi.* Al-Bilawi was taken to Camp Bucca, a large detention facility near Umm Qasr, a desolate stretch near the Persian

---

* Abu Musab al-Zarqawi was killed in an American air strike of his safe house near Baghdad on June 7, 2006, when F-16 fighter-bombers dropped two 500-pound bombs on his compound. Subsequent raids by Iraqi and coalition counterterrorist forces resulted in scores of AQI leaders arrested and a treasure trove of intelligence material about his network throughout the country.

Gulf coast south of the city of Basra. Thousands of jihadis and Baath Party insurgents were warehoused at the Camp Bucca theater internment facility, named after a New York Fire Department marshal killed in the September 11, 2001, attacks. It was in Camp Bucca where al-Bilawi met another one of Abu Musab al-Zarqawi's lieutenants, named Ibrahim Awwad Ibrahim Ali al-Badri al-Samarrai. Al-Samarrai was the founder of Jeish Ahl al-Sunnah al-Jamaah, a militant Islamic group that found shelter inside the beleaguered Sunni enclaves.[24] Al-Samarrai would soon become known to the world as Abu Bakr al-Baghdadi— the future emir of the soon-to-be-declared Islamic State.

At one point there were twenty-four thousand inmates crammed inside the tight confines of Camp Bucca. The prison quickly became an Ivy League university for future terrorists. Inmates honed their skills and sharpened their ideological and religious fanaticism behind the thick coils of razor wire; harsh interrogations by American and Iraqi staffers only hardened the prisoners and their violent resolve. A US Air Force security officer who commanded one of the facility's compounds said that it was obvious to anyone there that the detainees spent their years focused on two things: radicalization and revenge.[25]

The Camp Bucca theater internment facility was closed in 2009. Many of the prisoners were simply let go, released as part of an amnesty meant to rehabilitate hard-core militants by reintroducing them into Iraqi society. Those deemed to be too much of a security risk to reenter society were sent to other facilities throughout the country. Al-Bilawi spent four years within the walls of Abu Ghraib, the notorious maximum security near Baghdad. The other prisoners treated him with the reverence that one hundred years ago was reserved for a Turkish pasha.

On the night of June 21, 2013, several suicide car bombs were driven into the main gates of Abu Ghraib prison, rocking the compound. The explosions were followed by multidirectional automatic weapons fire; guard towers were destroyed by relentless barrages of RPG fire. Men clad in black fired as they praised God; they took up positions near the highway to the capital, fighting off security reinforcements sent from Baghdad as strike squads, wearing suicide vests, entered the prison on foot to help free the prisoners.[26]

The highly coordinated strike had been in the works for months, and high-ranking prisoners, many on death row, helped to plan the bold prison break. Over five hundred maximum-security prisoners vanished before dawn, including al-Bilawi.

The men who launched the daring raid on Abu Ghraib prison belonged to Abu Bakr al-Baghdadi's al-dowla al-islaamiyya fii-il-i'raaq wa-ash-shaam, the Islamic State in Iraq and the Levant, or ISIL; the entity would also become known as the Islamic State in Iraq and Syria, the Islamic State, or ISIS. In the blink of an eye, ISIS had morphed from a terrorist underground into a full-fledged army. Most Arabs mockingly called this latest homicidal band of holy warriors by the Arabic acronym of "Da'esh," when, if pronounced a certain way, had a myriad of un-Islamic connotations, including one that meant a shameful lack of dignity.[27] The legion marching across the Iraqi desert threatened to pull out the tongues of anyone caught using the insulting term; the threat was taken quite literally.*

ISIS was a uniquely Sunni phenomenon defined on a foundation of religious purity and medieval barbarity. A more diabolical and a far more ambitious offshoot of what was al-Qaeda in Iraq, ISIL men who flocked to the fight were fueled by rage and revenge against a whole host of enemies. Most of the bloodlust, though, was targeted at the hated Shiites and was driven by a thirst for religious justice.

The spill-off from the US-led invasion of Iraq, namely the entrenchment of Shiite leadership in Baghdad, was more than enough to enable this new Sunni group to flourish. The sav-

---

* Other derivations of the word "Da'esh" would later be interpreted as "bigots who impose their views on others."

agery reserved for Shiites caught inside the Islamic State advance across Iraq was alarming even by Iraqi standards. Islamic State fighters left a scorched-earth landscape of death and devastation in every town and village they captured. There were mass executions and beheadings for the Shiites and other nonbelievers trapped in the path of this new army's speedy advance, and the militants weaponized rape as a cruel tool of war.[28] Wherever the black banners of the Islamic State were hoisted atop rooftops, indicating that the terrain had been liberated by the pure and faithful, men, women and children lay dead, strewn about town squares or dumped in mass pits.

The new Islamist army came out of nowhere to bulldoze a path through the war-torn remnants of Iraq, gaining momentum—and new Sunni recruits—with every kilometer they seized. Fallujah fell in January 2014, after a pitched battle against the American-backed Iraqi Army. Over the next few months, the Islamic State swept across much of Anbar Province, which covered most of southwest and central Iraq—from the approaches to Baghdad all the way to the border with Syria. ISIS continued into Syria, which was already torn to shreds by its brutal civil war and ripe to become host to this new regional contagion. The line in the sand meant to stop the Islamic State's charge was Mosul.

As June approached, Lieutenant General Gharawi and Abu Abdulrahman al-Bilawi sat in their respective command posts reviewing maps, positioning men and preparing for the inevitable showdown. The summer sun burned without mercy, and the Islamic State violence that scorched the landscape was fast approaching from the east. The army of no more than fifteen hundred armed men moved west at frightening cadence and closed in on Mosul, Iraq's second largest city, in the final days of May 2014. The Islamic State's forces encroached on the city in the dust clouds that their Toyota Land Cruisers and Hilux pickup trucks kicked up on the sand-swept desert roads. The militants wore black fatigues and long beards with their upper lip left clean and shaven like the Prophet Mohammed. These men left a calling card of barbarity that made many flee in panic, desperately trying to stay one step ahead of the militants' speedy advance across the Levant (the term used to cover the historical confines of Lebanon, Syria and Israel/Palestine).

Iraqi commanders were fecklessly naïve to think that they could easily defeat the encroaching ragtag threat to Mosul. Over-confidence had been fatal to Iraqi commanders in their previous encounters with the Islamic militants.

Had Gharawi's intelligence chiefs bothered to gauge the pulse of the Sunnis and the other minorities living in Mosul, they would have found a malignant Sunni resentment metastasizing throughout the city. The Shiite-dominated government did everything in its power to make the non-Shiite communities feel threatened and persecuted. Taxes that were levied on non-Shiite businesses were exorbitant; there was an infestation of corruption and nepotism in every element of daily life for the city's multiethnic middle class. There were frequent knocks on the doors of Sunnis in the middle of the night by men in masks ostensibly representing the police. Fathers and husbands disappeared, never to be heard from again.

Restrictive checkpoints manned by Shiite soldiers dotted the two main roads leading in and out of the city. Sunnis and Kurds pulled over for inspection were almost always treated harshly. "Once a soldier looked at a traveler's ID card and saw his name, revealing his faith, the abuse could be as varied as a shakedown for cash to a beating or incarceration," Majd Holbi, a war correspondent in the region, remembered. "Sunni and Kurdish women were humiliated in front of male members of their family. People wanting to pass along were often encouraged to sing Shiite songs at gunpoint. People wanting to travel or do business never knew which one of the Shiite soldiers would snap and kill them for no reason."[29]

Hatred of the Shiite-led regime in Baghdad was tangible in

Mosul, yet the Iraqi National Intelligence Service, the American-designed successor to Saddam Hussein's dreaded Mukhabarat, was too busy using Mosul as a convenient ATM. The massive American intelligence presence in Iraq also failed to pick up on the discontent.

As al-Bilawi's forces closed in on Mosul, Gharawi positioned his forces for the inevitable showdown. Gharawi had two divisions—the Iraqi 2nd and 3rd—at his disposal, complete with the latest US-supplied M1A1 Abrams main battle tanks and scores of MRAP (Mine-Resistant Ambush-Protected) armored vehicles. There were close to sixty thousand men under Gharawi's command, including Iraqi special forces battalions that were American trained and had combat experience. Gharawi and his brigade commanders deployed their forces for a winner-take-all showdown at the gates of the city, but he was playing a losing hand. It was common for Iraqi servicemen to kick back as much as half of their wages to their commanding officers for the right not to show up at all.[30] Other soldiers and policemen sold their weapons for cash, for barter or for sex. Most of the American-supplied high-tech equipment was never removed from their watertight crates; other pieces of gear were already in the hands of ISIS.

Masoud Barzani, the president of semiautonomous Kurdistan Regional Government, knew just how dire Gharawi's position was; the Kurdish intelligence apparatus was spot-on in its infiltration of the Islamic militants. But Baghdad refused to listen. Prime Minister al-Maliki was confident that his American-trained, American-equipped and American-advised Iraqi Army was poised to defeat the Islamic State in one decisive, go-for-broke battle. Gharawi vowed that the Sunni militants would never raise their black flags inside Mosul. Meanwhile, the Islamic State was already in the city.

Mosul was the true treasure of the Islamic State's plans for Iraq and Syria, and al-Bilawi's sleeper cells were well entrenched inside the city just awaiting the orders to attack. Their weapons and explosives were hidden inside the homes of Sunni families throughout the western half of town. The Sunnis who assisted the militants weren't necessarily interested in the fanaticism of the armed men. "Mostly," Majd Holbi remembered, "the Sunnis just wanted to be left alone. They didn't want to be harassed, humiliated and harmed. Women were abused. The Shiite soldiers at the checkpoints terrorized them."[31] Gharawi had ordered the extrajudicial killings of Sunnis inside Mosul, leaving most with little choice but to take sides with the fanatics.

The heavy-handedness turned the city's Sunni population into dried tinder that al-Bilawi would soon ignite. The militant's takeover of virtually all organized crime in the city generated an overflowing war chest that funded al-Bilawi's plan to seize the city.[32]

The uninterrupted crackle of automatic weapons, like the onset of thunder, grew louder as Islamic State forces closed in on Mosul from the desert. Fires billowed in the distance. The smoke, sun and desert sands whipped up an eerie copper-colored cloud over the doomed city.

In late May, Gharawi's counterintelligence agents arrested seven Islamic State lieutenants. Their interrogations revealed that ISIS was planning its Mosul offensive in early June. Gharawi decided to launch a preemptive strike. On June 4, Iraqi police special operations units raided al-Bilawi's safe house and command center where al-Bilawi was running his Mosul attack. Rather than be taken alive by Iraqi forces, al-Bilawi blew himself up.

Gharawi's commandos returned to headquarters with a treasure trove of intelligence taken from al-Bilawi's home. The material was so voluminous that it had to be transported back to base in several large military kit bags. The haul included computers, hard drives, printouts and maps that detailed every element of the planned attack against Mosul. Gharawi informed

Baghdad that the much-feared assault on the city had been pre-empted by the successful commando raid and al-Bilawi's death. Later that evening, though, Islamic State artillery batteries let loose on Mosul's outer defenses. The shelling lasted close to two days. The battle for Mosul had begun.

On June 6, suicide car bombs softened up many of Gharawi's main defensive positions around the city. Gharawi tried to mobilize his army but only a fraction of the Iraqi security forces answered the call to arms. ISIS units moved swiftly through areas that had been predetermined and, in many cases, booby-trapped with IEDs by the many sleeper cells al-Bilawi had secreted in the city. The main attack commenced on June 6. There were supposed to be, as far as Baghdad was concerned, anyway, close to twenty-five thousand military and security personnel defending Mosul. The reality was catastrophically different. There were only forty soldiers at their posts and on duty in Mosul's Musherfa district, a gateway to the city, on the night of June 6.

The Islamic fighters attacked Iraqi positions with fast-moving Land Cruisers fitted with heavy machine guns mounted in the cargo hold. Assault squads, some armed with suicide vests, others armed with RPGs and small arms, overwhelmed Iraqi defenses in ferocious small-unit attacks. There was, in retrospect, little wonder why Gharawi's forces collapsed so quickly. The officers who bought their commissions, those who bothered to show up at all, were inept field commanders. Some of these commanders were mere cowards. When they ran, most of the men they led fled at the sight of the men with the long beards waving their black al-Qaeda flags and banners without ever firing a shot.[33]

The Islamic State attack was carried out by only several hundred men who never, in their wildest dreams, thought that they could defeat an entrenched Iraqi force so vastly superior in size and boasting state-of-the-art American-supplied hardware. The battle, though, was a call to arms for like-minded men throughout the area. ISIS fighters in Syria crossed the border from Raqqa to join in the victory. Once again Kurdish president Barzani offered fighters to bolster Mosul's defenses. Once again al-Maliki refused.

In the Sunni neighborhoods the bearded men waving the black flag were treated as liberators. For the Shiites, especially the security forces, the day of reckoning had arrived. Islamic State fighters captured—and then quickly executed—scores of Shiite soldiers and policemen. The prisoners were murdered in the most gruesome of manners: hanged, burned alive, crucified or beheaded. The headless corpses were displayed in neighborhood traffic circles for all to see. Some of the fighters decorated their Land Cruisers with the heads of the men they had killed. Women, including teenagers and young girls, were already being abducted off the streets. Sexual assaults and rapes were rampant. The fighters exacted special vengeance on Mosul.

Islamic State propaganda made a huge effort to publicize the idea that the ferocity of their fighters—men who came from all over the Levant and, ultimately, from all over the world—was due to their piety and their zealous sacrifice in creating the modern-day Caliphate. The real reason, as one former commander expressed years later, was for the money. "I don't have anyone, I'm poor," the captured jihadist stated. "I only have God. For me, it was about the money."[34]

The terrified residents of Mosul—Shiites as well as Sunnis who didn't agree with the fanaticism of the militants—didn't care about the motivation of the men carrying out atrocities before their eyes, and they didn't have to wait until the Land Cruisers and captured High Mobility Multipurpose Wheeled Vehicles (HMMWVs also known as HUMVEEs) sporting loudspeakers and black flags came to warn civilians that they were now citizens of the Caliphate before deciding to flee. The human exodus of Shiites, Kurds, Yazidis, Christians fleeing Mosul was epic in scale. Nearly half a million men, women and children left between June 6 and 9.

By June 9, Gharawi's hope to hold the city came to a crashing end when the influx of Syrian fighters and sleeper cells proved too much for the defenders. On June 10, reinforced ISIS units stormed the city's notorious Badoush prison and killed the guard force. The liberated prisoners were interrogated by ad hoc inquisitors to determine who was a Sunni, who was a Shiite, and who was a Kurd or a Yazidi. The prisoners were then trucked to the desert where, according to reports, an Afghan in Islamic State service ordered Sunnis to the left and everyone else to the right. The Afghan allegedly said, "If I find out that a Shia is among the Sunnis, I'm going to cut off his head with a sheet of metal." Over six hundred non-Sunni inmates were then systematically slaughtered in cold blood.[35] The Sunnis were given food and water. Many grabbed a Koran and an AK-47 and raced to the front lines to join in on the fight.

As the battle for Mosul raged, Gharawi and his men were forced into a final and indefensible position near an abandoned hotel in the heart of the city. ISIS struck the perimeter with a suicide truck bomb and then with swarms of attackers. Most of Gharawi's lieutenants fled in the darkened chaos across the Tigris River. Gharawi barely escaped capture and a ghastly end in a daring dash for safety. Sunni members of the Iraqi armed forces and police joined the ranks of the Islamic State. Everyone else ran for their lives. By June 10 it was all over. Mosul, Iraq's second-largest city, with vast gold and cash reserves and the oil wealth of a Persian Gulf emirate, fell to only a few hundred armed men of a terrorist army in ninety-six hours.

ISIS forces now had more than enough money and oil surplus to fund and fuel their march across the Middle East. The Islamic State captured hundreds of American-supplied armored vehicles and enough advanced arms and ammunition to arm a small nation; the jihadists captured over fifty howitzers and even top-tier American-made M1A1 Abrams tanks.

ISIS staged a military parade following the city's capture. Joyful militants, praising Allah as they fired their AK-47s wildly into the air, drove down the city's main thoroughfare in their captured tanks and blast-resistant armored vehicles. For the first time in history, a major terrorist group had metastasized into a nation-state that had natural resources, vast cash and oil reserves, and the means to field a conventional army that could seize large swaths of territory. It was the true nightmare scenario where terrorism and vast wealth intersected. The city was now the world's most important center of the global jihad.

ISIS moved on, advancing to points north and south of Mosul. Abu Bakr al-Baghdadi's army launched an offensive to seize more terrain and take advantage of hubris in Baghdad, the chaos in Syria and a calculation that the Western powers, led by the United States, were in no mood to embark on a new war in the Middle East.

Shortly after Mosul fell, a cryptic radio communiqué from Raqqa declared the establishment of the Islamic State, a caliphate, in the territories that the militants controlled in both Syria and Iraq. "Listen to your caliph and obey him. Support your state, which grows every day," Abu Mohammed al-Adnani, the group's spokesman and operations chief, proclaimed.[36] The group declared that it had removed "in the Levant" from its title. It would simply be known as the Islamic State.

Days later, the enigmatic and highly secretive al-Baghdadi emerged from the shadows to appear before Friday prayers in Mosul's legendary Great Mosque of al-Nuri, famous throughout the region because of its leaning minaret. Wearing black robes, the new caliph climbed the marble stairs toward the pulpit with a slow and confident stride and, in broad daylight, declared war on the world. "If you want what Allah has promised," al-Baghdadi declared, "wage Jihad for His sake."[37] His sermon-like proclamation was videotaped with dramatic direction worthy of the best Hollywood film crew. The propaganda value was enormous as the footage went viral. Al-Baghdadi's appearance in the mosque was the first—and last—time he'd ever be seen in public.

Foreign fighters flowed into this new rogue state from the four corners of the world. Many of these men were ripe for the recruitment—unemployed, uneducated, first-generation residents of a postcolonial reality—and drawn by the promise of a weapon, a bundle of cash and a wife. Some of the fighters believed in the religious promise of the Caliphate. Others were simply psychopaths and thugs. They used their weapons to terrorize other Sunnis who did not adhere to their brand of Wahhabi-like fundamentalism. They used their own perverse interpretations of their religion to murder and rape and loot the area of its wealth. Those unable to flee Mosul—Sunnis and Shiites alike—bore the brunt of this evil.

A new dark age had quickly descended on the Middle East.

3

The Line in the Sand

There is a big difference between fighting the cold war and fighting radical Islam. The rules have changed, and we haven't.

—John le Carré[38]

AS FAR AS those who had the power to call in an air strike or redirect the orbit of a satellite were concerned, the epicenter of the Middle East was found inside a compound of buildings in an eastern slice of the Jordanian capital of Amman called Wadi Seer. There were ten neighborhoods in Wadi Seer, and the headquarters for the General Intelligence Department, or GID, was based in the residential section known as Jandaweel.

There was extraordinarily heavy security along the hilly road leading to the main GID campus in Jandaweel, complete with checkpoints manned by heavily armed special operations troops in red berets and fortified posts that trained heavy machine guns on every approaching vehicle. The GID was all-powerful in the kingdom, and it was often referred to as the national central nervous system. It spied on threats both foreign and domestic and wielded a reputation for fierce efficiency. Because the GID monitored threats inside the kingdom, it was feared at home; to those who meant to undermine the country's stability, the mere mention of the Jandaweel neighborhood provoked a sense of great foreboding. Because it was very capable in monitoring Jordan's enemies throughout the Middle East nation-states and their armed proxies, the GID's reach was far and revered. It was the GID's role in the global war against terrorism, however, that made the intelligence service stand out as the West's—and the world's—frontline shield. Its relationship with the CIA has become particularly close.

When CIA officer Jack O'Connell traveled to Jordan in 1963 to establish functional intelligence ties with the kingdom, he met with King Hussein, then just twenty-eight years old. "I would like to meet with the head of the Intelligence Service to establish a working relationship," O'Connell requested. "You are looking at him," the Jordanian monarch responded. "I am the head of the Intelligence Service."[39]

The CIA's relationship with Jordan's intelligence service, known for a while simply as the mukhabarat, Arabic for "information gathering," was always considered to be a mutually beneficial one. The countries established a close-knit operational alliance that transcended administrations, political parties and the men who ran both organizations. The United States invested heavily in Jordan's armed forces and its intelligence apparatus. King Hussein's spies acted as go-betweens, and as a force to ensure moderation in the region—intelligence was a commodity worth more than oil in the volatile Middle East, and knowing what was happening, who was a friend and who could potentially be a threat was essential in keeping the region's countries stable. Jordan's intelligence officers were respected for their knowledge and insight, and they were used by King Hussein as trusted emissaries and as ambassadors who enhanced Jordan's standing in the world—with both its friends and its adversaries.

The Daera Elmukhabarat al Ameh, the Arabic name for the

GID, was created in 1964 by royal decree. Act 24 of the nation's parliament outlined the duties of this nascent force to safeguard the security of the Hashemite Kingdom of Jordan domestically and abroad by means of carrying out necessary intelligence.[40] The first years were, indeed, challenging for the nascent spy service: Palestinian guerrillas operated on the West Bank as a base of operations against Israel, Jordan was at war with the Jewish State, and forces inside the Arab world—especially from Egypt and Syria—were determined to undermine its stability and the rule of King Hussein. Harsh tactics were needed and indeed employed to keep the country from being thrust into civil war or domination by Jordan's Arab neighbors—some oil rich and clients of the West and others who were squarely in the Soviet orbit—who viewed the kingdom as a target rather a regional player.

The June 1967 Six-Day War and the sudden and surprising Israeli victory changed the landscape forever. Jordan lost the West Bank and Jerusalem to Israeli forces, while at the same time receiving close to a million Palestinian refugees who streamed across the Jordan River. Jordan suddenly found itself a front-line nation in the global counterterrorist campaigns and wars that would follow.

On September 6, 1970, Palestinian terrorists hijacked two Western airliners over Europe and forced the aircraft to land at Dawson's Field near Amman. Days later, after the hostages were removed from the planes, the terrorists blew up the airliners in a powerful message captured by the world's media. The Palestinian operation was a direct challenge to Jordan and tensions, simmering for several years, erupted. The GID found itself on the front lines of a brutal civil war pitting Jordan's armed forces against a myriad of Palestinian groups. The fighting raged primarily around the capital city, and it was fierce, with both sides giving no quarter. Syria invaded Jordan to save the outmatched Palestinian forces, but the incursion was beaten back. It took a month until the Palestinian groups were forced into Lebanon; they would set up new bases of operations from which to attack Israel and the West.

The Palestinian liberation fronts never forgave King Hussein for the war it called ailul al-aswad, or Black September. A covert terror force by the same name debuted in November 1971 when it assassinated Wasfi Tal, the Jordanian premier, as he left an Arab League conference in Cairo. The assailants, eager to show the world their ferocious abandon, lapped up the slain man's blood while it flowed from his wounds.[41] Jordan's fight against Black September would define the GID for generations to come. The intelligence service became expert in infiltrating Black September's networks in ways that the other espionage agencies that worked in the region, including the American CIA and Israel's Mossad, couldn't compete with.

The GID's HUMINT, or human intelligence, capabilities separated it from most other Middle Eastern intelligence operations.[42] The country's minorities flocked to the ranks of the security forces, and they especially excelled inside the intelligence service: Circassians and Chechens, ethnic groups that fled persecution and found both a haven and land inside the Ottoman-controlled area, excelled as case officers and commanders; the Druze and Christians also flocked to the security force. Bedouins, the backbone of Jordan and many with tribal ties throughout the region, made up most of the spies and spymasters inside the organization. The country's tribal culture and clannish bond of the minority groups valued loyalty over short-term expediency.

The minorities and the Bedouins understood the mind-set of the men and organizations they targeted. "The GID didn't have shipping containers full of cash to bribe sources," a retired psychologist from the spy service explained, "but we were born knowing the mind-set of one another, speaking the language and I'm not only referring to the Arabic. We were trained from kids to speak and hear the nuance of the tribes and of the minorities and to understand what made them tick and what caused them to turn to violence. As a nation that absorbed ethnic groups rather than displaced them, we understood what fears these people possessed, as well as to where their comfort zones and those elements—fear and reward—were useful tools in intelligence campaigns. We also knew the trip wires of the Arab psyche—when and how to apply pressure and when and how to administer reward. All the money in the world can't buy these priceless skills," he added, "and all the academic studies can't teach it."[43]

The service's exceptional HUMINT capabilities allowed the GID, as John le Carré so aptly described, to sit at the "high table" in the global intelligence fraternity alongside agencies such as the CIA, Britain's MI6 and others.[44]

The GID's stock rose exponentially following the September 11, 2001, attacks against the United States. The GID was, perhaps, the only reliable Arab intelligence service that had sources inside al-Qaeda and that shared their invaluable assets with the Western services. Jordanians had joined Osama bin Laden's war against the Soviet Union, and some had remained in Afghanistan to fight on behalf of al-Qaeda; others returned to Jordan where they reentered civilian life, though still harboring close links to the men in the caves who would fight the international coalition following 9/11. "They know the bad guy's culture, his associates, and more [than anyone] about the network to which he belongs," a former CIA officer was quoted as saying in the *Washington Post*. "[The GID] has unrivaled 'expertise with radicalized militant groups and Shia/Sunni culture.'"[45] When the *Washington Post*'s noted columnist David Ignatius asked CIA director George Tenet which intelligence service had been most successful in penetrating al-Qaeda, the answer was "the Jordanians. They're superstars!"[46]

Together with the CIA, GID officers ran agents in Afghanistan to get closer to the impenetrable ring of security that protected bin Laden and his inner command roundtable. In one such operation in 2010, GID officer Sharif Ali bin Zeid was killed, alongside seven CIA officers, at Camp Chapman in Khost Province. The allied intelligence team was targeted when a Jordanian double agent the GID officer was running turned out to be a triple agent working for al-Qaeda and blew himself up at the clandestine meet. Sharif Ali bin Zeid was a cousin of King Abdullah.

It was in Iraq, though, that the GID's reach into the radical networks operating in the region proved invaluable to the allied cause. The leader of al-Qaeda in Iraq, Ahmad Fadeel al-Nazal al-Khalayleh, soon to be known to the world by his nom de guerre of Abu Musab al-Zarqawi, was himself a Jordanian.

Born in 1966 in the industrial town of Zarqa, northeast of Amman, al-Khalayleh was a thug, a petty thief, a drug dealer and a pimp. The GID had an extensive file on the man whom most inside the intelligence service, a senior counterterrorist officer reflected, believed would spend his life behind bars.[47] He had been radicalized in prison and in 1989 he left Jordan for Afghanistan, hoping to take part in the Islamic Army's war against the Soviet invaders, though he arrived at the tail end of the conflict when the war's outcome had already been sealed.[48] He spent five years in Afghanistan and, reportedly, the fundamentalist terrorist recruiting facilities in Pakistan's Peshawar Province, though he returned home in 1994. The GID discovered that al-Khalayleh was stockpiling explosives and weapons in his home and he was imprisoned for sedition and terrorist offenses; he was the founder of a group, known as the "Syrian Division," that wanted to continue the bin Laden–style global war inside Jordan.[49] He served his sentence in the notorious Swaqa prison, some sixty miles southeast of Amman in a desolate stretch of the desert. As a veteran of bin Laden's first war against a superpower, al-Khalayleh was a celebrity behind bars, and he became a prison commander of like-minded men. Both the GID and the prison services were aware of al-Khalayleh's past as a violent predatory felon and recidivist offender, but his fundamentalist views of Islam metastasized even further behind bars.

In 1999 al-Khalayleh was released from prison in a general amnesty. Some in the GID expressed grave concerns over releasing someone like al-Khalayleh, a man that veteran GID agents, even his jailers, believed was beyond rehabilitation.[50] The benevolence worked for some but backfired miserably in the case of al-Khalayleh. Once freed, he traveled to Peshawar, Pakistan, in the hope of establishing a terror cell made up of Jordanian Islamists, including those freed with him in the amnesty, and to launch a series of high-profile and bloody terrorist attacks inside the kingdom on New Year's Day 2000. One of the plots, designed to transpire around the millennium celebrations, was for suicide truck bombers to blow up the fully booked Radisson SAS hotel in Amman, as well as tourist sites holy to Christians along the Israeli border. The GID intercepted mobile-phone chatter between the plotters and their operatives in Jordan, and twenty-eight men were arrested before they could execute the attack; several of the terrorists were sentenced to death.

A warrant was issued for al-Khalayleh's arrest but he entered Pakistan and then crossed the mountainous frontier into Afghanistan and hooked up with Osama bin Laden and the al-Qaeda hierarchy. In Afghanistan, al-Khalayleh took on a nom de guerre, Abu Musab al-Zarqawi, a salute to the gritty town that created him, and established a training facility near Khost, where his lieutenants focused on explosives and poison gases. Zarqawi's goal was to create an Islamist army that could create massive unrest in Jordan. He was in Iran when the airplanes hit the World Trade Center and the Pentagon, then returned to Afghanistan to help bin Laden fight off the American-led military invasion. Zarqawi ultimately returned to Iran, and from there moved to Iraq and set up sleeper cells inside the Iraqi capital under the eyes of Saddam Hussein.[51] As the Middle East began to brace for war, expecting the United States to invade Iraq,

Zarqawi launched his first strike as the head of what would become known as al-Qaeda in Iraq. On October 28, 2002, gunmen assassinated Laurence Foley, a senior American diplomat, as he left his Amman home. Abu Musab al-Zarqawi became the GID's top target—wanted dead or alive.

Following the US invasion and the subsequent dismissal of the entire Iraqi civil service—from mailmen to military officers—Zarqawi and his lieutenants launched a lethal blitzkrieg of suicide bombings and IED attacks, killing scores of American troops and thousands of Iraqi civilians. Zarqawi's network also targeted Jordan. In 2004 an al-Qaeda in Iraq cell attempted to blow up GID headquarters and attack a crowded soccer stadium with poison gas, but Jordanian spies uncovered the plot and counterterrorist forces terminated the cell. A year later, though, Zarqawi succeeded in carrying out a lethal strike in the Jordanian capital when three separate cells launched suicide strikes against hotels in the city. Sixty people were killed in the coordinated attacks and scores more wounded.

Zarqawi's reign of terror made him the most wanted terrorist in the world next to bin Laden. A US Special Forces strike team, known as Task Force 145, hunted Zarqawi for close to three years, but the thief and pimp turned master terrorist became very successful in recruiting from the tribal areas of Anbar Province in southern Iraq close to the Jordanian border; he was protected by the Sunni clans in the desert. But the GID was very active in Anbar, as well, and the Jordanian spies were expert in linking the cousins and in-laws who were connected to Zarqawi into a web of tantalizing and accurate intelligence. The information was vital in helping Task Force 145 pinpoint a safe house near Baghdad where Zarqawi was meeting with some of his lieutenants. On June 7, 2006, two US Air Force F-16s bombed Zarqawi's hideout, killing the elusive terror chieftain and several of his lieutenants.

The GID became an intrinsic element in the allied effort to

stabilize post-Zarqawi Iraq. The GID possessed access to the tribal elders, and the intelligence they were able to glean from these blood ties was beyond the scope of what the CIA could collect with bundles of cash, promises of green cards and all the satellite eavesdropping that money could buy. Infiltrating a network, using bonds that cannot be corrupted by cash, was where the GID was in a league all by itself. It has been written that, because of Iraq, the GID became the CIA's closest intelligence partner after Britain's MI6. "The GID," an Israeli intelligence veteran commented, "was punching above its weight and impressing the premier members of the league of services who had more money, more men and more global reach. They rose to the challenge and then some."[52] Charles Faddis, a former CIA operations officer who headed the WMD counterterrorism unit, wrote about the CIA relationship with the GID, and stated, "In many ways, this relationship has become the template against which all others are measured."[53]

The CIA station in Baghdad was the largest in the Middle East (Cairo was second), but when American forces withdrew from Iraq the CIA's intelligence-gathering efforts were detached and, in many cases, no longer coming from reliable human sources. Now they relied upon technological means of gathering and interpreting intelligence, conducted from behind a desk and in front of a computer monitor instead of frontline eyes-on target vantage points. The GID was the only allied intelligence service left in Iraq with an active ground game. The intelligence that the Jordanians relayed back to CIA headquarters in Langley, Virginia, were troubling. The United States was so heavily invested in the Iraqi regime and its security forces that it became commonplace, almost second nature, for many in the CIA, inside the Pentagon and, indeed, inside the White House to accept the stark realities of what was happening inside Iraq and how that could affect what was happening in Syria from an intelligence contingent in country that was often learning of developments from a distance or through secondhand sources. "The more money and arms went into Iraq," one Jordanian intelligence officer commented, "the more many in Washington believed that everything would be OK."[54] King Abdullah read the same GID reports that were given to Langley. The assessment from the Iraqi theater of operations was grim.

General Faisal Shobaki was appointed GID Director in 2011 and took command of an intelligence service grappling with the Arab Spring and, later, the emergence of the Islamic State. Shobaki was a thirty-year veteran of security service to Jordan; his appearance was elegant and academic, and he looked the part of an economics professor merged with George Smiley. All GID Directors were given the title of al-Pasha, or "the Pasha." The designation was a sign of absolute reverence and was in deference to the country's Ottoman history; since Arabic doesn't have a "P" sound, the designation was always pronounced *Basha*. From experience he knew that what was brewing to Jordan's north, along with the resurgence of fundamentalist forces in Egypt and Libya, Africa and Somalia, and, of course, Iraq and Syria, would ultimately embroil much of the region in a bloodier and far riskier global war on terror—a conflict that would be harder to control and contain than the previous decade's worth of campaigns against al-Qaeda.

Faisal Pasha made frequent trips to Washington, DC, to meet with CIA director John Brennan and others in the US intelligence community to coordinate responses to the deteriorating situation in Syria and Iraq throughout 2013 and 2014. The two directors were lifers in their respective agencies.

John O. Brennan grew up in the shadows of New York City in an upper-middle-class bedroom community across the Hudson River in New Jersey. He joined the CIA in 1980 after earning a master's degree with a focus on government from the University of Texas in Austin; he focused on the Middle East and became fluent in Arabic after spending a year as an exchange student at the American University in Cairo. Brennan began his intelligence career as an analyst, though true to his Irish roots he was parochial and ambitious and advanced up the chain of command quickly. He landed choice postings requiring political savvy, including personally delivering the daily intelligence brief to President Bill Clinton. He was promoted to chief of station in Riyadh, Kingdom of Saudi Arabia, when Hezbollah launched a suicide truck bombing of the US Air Force barracks at Khobar Towers. He later was appointed the chief of staff to CIA director George Tenet in 1999. After 9/11 he went on to head the Terrorist Threat Integration Center (TTIC).[55] Following a brief leap into the corporate world, he returned to government service, working in the Department of Homeland Security before being appointed by President Obama to serve as CIA director. George Tenet, his former boss, described him as being "smart as hell and tough as nails."[56]

Now, Shobaki urged Brennan that action was needed from the American side to prevent a calamity in Iraq. King Abdullah intervened, as well, upping the ante to lobby friendly lawmakers such as John McCain to act and preempt the disasters that were imminent. There was only silence. And then, to everyone's surprise, Mosul fell. It was, as one diplomat in Jordan commented, "a true oh shit moment."[57]

The Islamic State's black flags flying high over the mosques of Mosul was *the* true nightmare scenario that capitals in the region—and around the world—had always feared but, apparently, had never prepared for. President Barack Obama's White House was caught completely by surprise by the developments in Iraq. The Obama administration was already reeling from the miscalculated declaration that a red line would be crossed if Syrian dictator Bashar al-Assad would deploy chemical weapons in the civil war.[58] Syrian forces had, of course, gone on to use chemical weapons against its citizens, and the absence of a

US military response was a stark example of a lack of American commitment to the Middle East. The American president had built a foreign policy around disengaging the US military from the Middle East, and now he faced a region in dire crisis screaming for US leadership and military action.

Of course, the decision makers were left in the dark about the Islamic State, because after the withdrawal of American forces from Iraq, monitoring events on the ground without enough eyes and ears on target had left them with a detached and unfocused assessment of the frontline realities. The Central Intelligence Agency had also been caught off guard. CIA Station Baghdad was the largest in the Middle East but the office, as one intelligence agent* in the region surmised, "had begun to believe the mistruths and overestimations of its client, rather than skeptical supervisors of the information they were sending back to headquarters."[59]

The Pentagon was also guilty in not realizing that the sands beneath its Middle Eastern foundations were shifting.[60] Even though the Islamic State's march across Iraq had taken six months, the Department of Defense had not prepared any decisive contingency planning to react to the fall of Mosul. General Dempsey, the chairman of the Joint Chiefs of Staff, interviewed for the public television series *Frontline*, admitted to the shortcomings when he said, "There were several things that surprised us about ISIL [ISIS]. The degree to which they were able to form their own coalition, both inside of Syria and inside of northwestern Iraq; the military capability that they exhibited; the collapse of the Iraqi Security Forces… Yeah, in those initial days, there were a few surprises."[61]

The Obama administration was not the only one that failed to see that Iraq, after Syria, was catapulting into conflict and collapse. NATO hoped it could avoid Syria and Iraq 3.0, a third full-scale war in the area, in less than thirty years, at all costs.

---

* The officer has requested that his/her identity be withheld.

Wars in Afghanistan and Iraq, and smaller ones in Yemen and
the Horn of Africa, had exhausted the political will of Western
leaders to once again engage in Middle Eastern conflicts. President Obama had infamously declared that the United States
would respond militarily should a "red line" be crossed if Syria
introduced chemical weapons into the civil war. But when the
news broke on August 21, 2013, that Bashar al-Assad had ordered his forces to bomb a suburb of Damascus with sarin gas,
an attack that resulted in the deaths of over one thousand civilians, the Obama administration hemmed and hawed.[62] The
British were also reluctant to go to war again. Parliament voted
285–272 against joining the proposed US-led air strikes against
the Assad regime; the winning side of the debate, filled with
rancor and historical "I told you so's," was that Great Britain
should not follow the United States to war in the Middle East
again just as it had so erroneously done when Prime Minister
Tony Blair followed President George W. Bush into Iraq a decade earlier.[63] Now that Syria and Iraq were part of the same
conflict, turning a blind eye had been viewed as a safer course
of action than committing to yet another lose–lose fight.

Still, the irrefutable fact remained that the emergence of the
Islamic State had caught everyone by surprise inside the capitals
of the Middle East, as well. From Baghdad to Beirut, Tunis to
Tel Aviv, the gatekeepers had been asleep at the switch.

The Islamic State was emboldened by the collective shock
and awe it had inflicted on the regional powers, but this time
the United States responded quickly. Three weeks after the fall
of Mosul, the US Army Central Command (ARCENT) received the green light to conduct military operations inside
Iraq to stem the advance of Islamic State forces; it was the first
US military operational headquarters in the country since the
2011 withdrawal. Air strikes against Islamic State targets began
in August. Ground missions, including the use of special oper-

ations forces, were launched to recapture the Mosul and Haditha dams, as well as to assist Yazidi refugees trapped on Sinjar Mountain who fled in advance of the Islamic State and the inevitable murders, rapes and enslavement.[64]

Despite these actions, the ARCENT effort was lackluster and ineffective. Without a firm commitment of boots on the ground and sorties that could pummel the expanding Islamic State day and night, the Caliphate would grow and become more powerful. Former Mossad director Meir Dagan articulated just how potent a foe the Islamic State was when he addressed students at the LBJ School of Public Affairs, at the University of Texas, in Austin, in 2014. "The Islamic State group has been able to obtain money by robbing a bank of $500 million, taking control of oil fields worth $3 million to $6 million per day, kidnapping foreigners and receiving donations," Dagan explained. "Unfortunately, with this money, they are able to pay [for fighters] and they are functioning in a way like a state. Not only do they [support] fighters, but they have a great interest to recruit people from Western countries. They do not have to recruit many people. They must recruit a number that will serve purposes enough for the future. They understand, at some point in time, they are going to invade the west."[65]

The fall of Mosul made many world leaders feel as if their countries were next—whether they ruled nations that bordered the Islamic State or if thousands of their citizens had flocked to Syria to join the jihad.

The North Atlantic Treaty Organization's September 2014 summit was held at the historic Celtic Manor Resort, in the quaint town of Newport, in Wales. Sixty-one heads of state participated in the two-day policy-setting conference, and each traveled with a legion of foreign ministers, defense secretaries, ambassadors, generals and colonels, aides and staffers, not to mention the enormous security details such high-profile movements warrant. These dignitaries warranted great media attention, and there were over fifteen hundred accredited journalists assigned to cover the event, plus several hundred stringers, freelancers and other hangers-on. And, of course, any large assembly of power that convenes on matters of global security attracts the top echelons from the defense industry, each eager to sell the latest fighter jets, anti-ship missiles and light tactical vehicles. The head groundskeeper at Celtic Manor was furious that the organizers of the event allowed tanks and aircraft to be displayed on his precious links and on the grass he nurtured.

When the summit was planned the year before, NATO's winding down of the war in Afghanistan was slated to consume much of the agenda. But there were also planned working groups on China's expansion into the South China Sea, and other topics of discussion ranged from the Ebola crisis to how the alliance responds to global calamities. But the world got in the way of the original talking points. NATO leaders were determined to come up with some sort of concrete military response to thwart further Russian land grabs and involvement in Ukraine.[66] And Syria and Iraq dominated the policy speeches and the backroom meetings. The declaration of the Islamic State in Mosul's main mosque two months earlier presented a dire threat to NATO's interests in the Middle East. Containing the Islamic State and destroying it warranted swift and decisive military action. There was a sense of urgency at the summit; some felt that the meeting was historic. US Navy Admiral (Ret.) James Stavridis, a former NATO commander, commented that the gathering was "clearly the most important one since the fall of the Berlin Wall because of the clear level of multi-crises."[67]

Jordan was the only nonmember state invited to participate in the discussions that shared a border with the Islamic State. King Abdullah wanted two things from the NATO bosses— help and action. Jordan's national resources were overwhelmed by the influx of Syrian refugees crossing the border and international contributions from the United Nations and the International Red Cross were barely enough. The help that Jordan sought was financial and a commitment to military action. In closed-door meetings at the palatial Welsh resort, the Jordanian monarch pressed his counterparts in the alliance to initiate decisive military action against the Islamic State. The king's intelligence specialists provided in-depth detail as to the intentions of the Islamic State war councils and how Jordan was the next target in their sights. The intelligence came from firsthand accounts from GID human assets inside Iraq and Syria. The raw and unpolished intelligence provided chilling evidence of the Islamic State's capabilities and intentions. And all indications were the Islamic State had every intention of expanding the conflict—to Jordan.

Any attempt by the Islamic State to challenge King Abdullah's rule had catastrophic potential, especially considering that close to two thousand Jordanians had left their homes to fight alongside the fundamentalists. Their return to Jordan, possibly with the aim of expanding the Caliphate, was a dire threat to Jordanian internal security. According to a poll by the Center for Strategic Studies at the University of Jordan conducted in 2014, it is believed that over 10 percent of Jordan's 6.5 million inhabitants viewed ISIS and its ideology favorably.[68]

The jihadists weren't recruited from the poor and hopeless; those individuals were harder to recruit because faith was always easier when there was food in their bellies and a job to go to. Most of the recruits to the fight in Syria and to al-Baghdadi's army were middle-class kids who had university degrees. These young men were close to achieving success but found themselves to be overly educated and underemployed. The fiery sermons, available online and through phone apps, were designed to recruit these impressionable and angry young men to take up arms. "These young men were volatile, lost in a purgatory of sorts between fulfilling their potential and becoming a success and being stuck with a degree that suddenly didn't bring the reward it promised and little hope of retaining their foothold in a middle-class world with all the rewards that this brought, such as a wife, a home, a car, etc. The fundamentalist message appealed to those who sought the reason behind their inability to grab the rungs of success."[69]

The larger the Islamic State landmass grew in size, the richer it became, and the easier it was to recruit more middle-class jobless college-educated sons in the kingdom. That equation described the situation in Egypt, in Morocco and elsewhere in the Arab world—especially in Tunisia, the birthplace of the Arab Spring and the largest talent pool for foreign fighters.

President Obama and Britain's Prime Minister David Cameron understood that they couldn't let Jordan become Iraq or Syria. The country was stable and pro-Western, and moderate— it was a pillar of American and NATO foreign policy in the Middle East. The Jordanian Armed Forces, or JAF, were among the best in the Arab world. Most important, as far as American and British defense officials were concerned, Jordan was a trip wire manning a necessary vigil over the numerous trip wires that kept the day-to-day realities of catastrophic terror away from the streets of London or New York City. Jordan's security was also defined as the West's security. Any threat to the status quo risked a wider and far bloodier conflict in the Middle East. Behind closed doors King Abdullah made persuasive arguments with the NATO heads of state that the time to hit the Islamic State was now and with decisive force.

NATO did act. The NATO chiefs released a statement affirming that the "so-called Islamic State of Iraq and the Levant poses a grave threat to the Iraqi people, to the Syrian people, to the wider region, and to our nations. We are outraged by ISIL's recent barbaric attacks against all civilian populations, [in particular] the systematic and deliberate targeting of entire religious and ethnic communities. We condemn in the strongest terms

ISIL's violent and cowardly acts. If the security of any Ally is threatened, we will not hesitate to take all necessary steps to ensure our collective defense."[70]

President Barack Obama and British prime minister David Cameron went further by penning a joint opinion piece in the *London Times*, where they wrote, "We will not waver in our determination to confront" the militant group known as the Islamic State in Iraq and Syria, or ISIS. "If terrorists think we will weaken in the face of their threats they could not be more wrong."[71] Turning those words into action would take time, though. There was an international coalition, one not seen since the 1990–1991 Gulf War, that had to be cobbled together. There were base-of-operation agreements to be coordinated, decisions made by over a dozen nations as to which forces to deploy and what resources and rules of engagement to issue. There were also political ramifications to consider. Middle Eastern history was full of efforts of noble intent that disintegrated into full-blown carnage.

The NATO ministers' push for action was nothing more than a neat press release without decisive movement from CENT-COM, the US Central Command, headquartered out of Mac-Dill Air Force Base* in Tampa, Florida. The US Department of Defense established CENTCOM in 1983 as a theater-level command meant to meet rapid deployment commitments for an area of responsibility that included western and central Asia and North Africa. America's Middle Eastern wars—virtually all the campaigns against terrorism—were run out of the command's massive complex. On October 17, 2014, the Department of Defense formally established Combined Joint Task Force–Operation Inherent Resolve (CJTF-OIR) to formalize and or-

---

* The US Special Operations Command, encompassing the US Army Special Operations Command (USASOC), the US Air Force Special Operations Command (USAFSOC), the Marine Corps Special Operations Command (MARSOC) and the Naval Special Warfare Command (NSW), was also based out of MacDill Air Force Base.

ganize military actions against the rising threat posed by the Islamic State in Iraq and Syria.

Numerous flags were represented in the international alliance declared to defeat the Islamic State. Great Britain, Canada, France, Germany, the Netherlands* and other members of NATO pledged forces; aerial assets from these nations were rushed to forward-operating air bases throughout the Middle East.

There was a great deal of bewilderment at the start of the campaign. The battlefield was fluid, and the ROEs, the rules of engagement, were not designed for strikes against a ragtag enemy that used the hapless civilians around them as human shields. Territory sometimes changed hands several times in an afternoon, making the targeting aspect of ground support a challenge. The fighting was in Syria and in Iraq, yet assets were spread all around the Middle East—from US aircraft carriers in both the Mediterranean and the Persian Gulf to fighter-bombers flying from Jordan, Turkey, Qatar, Kuwait and elsewhere. The intelligence could change dramatically from the time the targets had been assigned to a sortie's launch and arrival over the target. Often coalition commanders would complain that they simply did not know where the front lines were. There was concern among some leaders that the strikes would be "willy-nilly" against a small village and that innocents or friendly forces would be hit, all but certainly guaranteeing new recruits to the Caliphate. Discussions of operational strategies were sometimes conducted in person and often via secure teleconference. King Abdullah offered the perplexed generals and diplomats an easy solution to unraveling the mystery. "You want to know where

---

*   Each nation came up with its own code name for the mission against the Islamic State. The British military referred to the campaign as Operation Shader; the French called it Opération Chammal (the French word for a northwesterly wind that blows over Iraq and the Levant); the Canadians called it Operation Impact; the Australians called it Operation Okra.

ISIS is? Fifty yards west of where the [Iraqi] Kurds, that's where ISIS is!"[72]

Initially the campaign was marked by inaction and hesitation. US commanders in Iraq told congressional delegations visiting the area that they needed six months to properly deploy special operations units to the front lines.

Unlike when Moses dispatched his spies to Canaan and they reported back that the land was flush with milk and honey, the intelligence operatives from many nations that crossed the Jordanian border into Syria to lay the groundwork for the Inherent Resolve campaign did not see a land blessed with rivers of flowing honey and sumptuous fruit hanging from every tree. Perhaps that was the case in the Syria of old. But not the country in 2014, which had already endured three years of sectarian civil war. Syria was a nation in name only, a series of bombed-out buildings with a ruptured landscape and a displaced and decimated people. The fighting was nationwide; few were exempt from the slaughter. The coalition spies returning from Syria did not come back from the desert bringing their Moses fruits to taste. The spies were lucky enough if they made it back to friendly lines alive.

Satellites and drones were invaluable resources to target vital Islamic State installations and leadership from various altitudes, but the spies were needed to fill in the blanks. The Islamic State advance through areas of Syria was rapid. Islamic State commanders were wily and fluid thinking: a weapons warehouse one day was a shelter the next. Collateral damage where innocent civilians were killed became fodder for the terrorists' slick and highly effective online propaganda apparatus. Not all targets were visible from space, as well. Some could be identified only by a man sitting inside a beat-up Peugeot, smoking a cigarette, while he watched and reported the comings and goings from behind a pair of mirrored Ray-Bans.

Faisal Pasha was busy. As the civil war and the emergence of the Islamic State blended challenges in Iraq and Syria into one mess, the GID mission included intelligence gathering—namely, spying on Jordanian volunteers that were now fighting for the Islamic State and Islamic State military chieftains eyeing the kingdom as the Caliphate's next prize. The Inherent Resolve coalition changed the scope of the intelligence service's mission to include proactive targeting.

The GID wasn't the only service to set up encampments along Jordan's border with Syria. The Americans were well represented on the desolate dusty roads along the border that were once used solely by military patrols and smugglers. The border area was a haven for mysterious types representing mysterious causes.

One of the coalition's true nightmare scenarios was for an intelligence officer to be captured by the Islamic State, so the spies crossed into Syria protected by a force of shooters. Most of the crossings were at night. Nighttime along the frontier was filled with the roar of supersonic fighter-bombers flying overhead. The aerial offensive was relentless. The air strikes against the Islamic State were launched twenty-four hours a day.

4

Send in the Vipers...

Offense is the essence of air power.
— **General H. H. "Hap" Arnold, USAAF**[73]

THE MEN APPOINTED by US secretaries of defense to helm CENTCOM were always soldiers who marched toward greatness. The most famous of all the generals to lead the command was Norman Schwarzkopf. Thanks to his live television news conferences during Operation Desert Storm, "Stormin' Norman" became the first general to also be a media superstar. General Anthony Zinni was nicknamed "the Godfather" and became President George W. Bush's special envoy to the Palestinian Authority and Israel in the effort to end the suicide bombings of the Second Intifada. General Martin Dempsey would go on to be named the chairman of the Joint Chiefs of Staff. General David Petraeus, the architect of the surge in Iraq and a master in the machinations of asymmetrical warfare, went on to become the director of the CIA, and Marine Corps general James "Mad Dog" Mattis would become a revered secretary of defense. Mattis served as CENTCOM commanding officer from 2010 until 2013—during the Arab Spring and the eruption of civil war in Syria. He was replaced by General Lloyd Austin in 2014.

Austin, a West Point graduate, was the first African American to lead CENTCOM. He had spent most of his career as a combat infantry officer. He led the 10th Mountain Division (Light Infantry) in Afghanistan during some of the most hellacious combat with al-Qaeda and Taliban forces. He held various senior level commands in Iraq, as well. He was known as a charismatic commander who was cool and decisive under fire. While he didn't have the flash of a Schwarzkopf or the flare of Petraeus, he had a knack for coordinating multinational forces as part of a larger coalition. General Austin oversaw Inherent Resolve.

The international coalition against the Islamic State was a hodgepodge of commands, languages, missions, and diverse and sometimes converging political and regional objectives. Turkey, a NATO member, initially wanted the ouster of the Syrian strongman, with its forces engaging Syrian government targets, often near and in opposition to Inherent Resolve objectives. Troubling reports indicated that Turkey's vaunted National Intelligence Service, the MIT (the Millî İstihbarat Teşkilatı) was allowing foreign volunteers into Syria; thousands crossed into Syria from Turkey from locations around the world—particularly from Western Europe.

Some of the European nations in the coalition wanted to participate in the campaign just enough to show the voters back home that they were taking a proactive role in the war against the Islamic State without exposing their forces or their political standings to too much of the risk. The last thing that any of the European leaders wanted was an errant bomb falling on an orphanage or a pilot being shot down and captured. It was essential that ground force contributions be minimal, and primarily special operations units acted as training cadres and forward observers.

The scope—and spread—of the coalition required a Herculean effort to administer and coordinate the basic command, control and communications of military operations. Allied aircraft were based in several countries that were separated by thousands of miles and unfriendly skies. Fighter-bombers taking off from the sprawling al-Udeid Air Base in Qatar might have refueling support taking off from Incirlik Air Base in Turkey; combat search and rescue (CSAR) capabilities to provide emergency response should a pilot be forced to eject over enemy territory could be based in Jordan or Iraq; sometimes, the CSAR forces were offshore, in international waters, on the flight deck of a US Navy nuclear-powered aircraft carrier. The resources of the coalition were spread thin and were represented by different languages, different SOPs (standard operating procedures), different operational protocols and different equipment. The coalition's catchy tagline of "One Mission, Many Nations" was accurate only insofar as the many nations involved.

Technology connected each one of the command's components. General Allen was not required to leave his air-conditioned quarters at Inherent Resolve headquarters in Camp Arifjan in Kuwait, several miles south of the capital of Kuwait City, to brief soldiers and airmen about to head into battle. Video teleconferencing (VTC) made it possible for an American general or a Canadian colonel to provide the mission A-to-Zs to Saudi Typhoon pilots from the No. 3 Squadron at King Fahad Air Base in Taif all from behind encrypted multiuser links. Highly classified information could be downloaded into the shared meetings separated by miles and times zones. There was a surreal disconnect to this type of warfare where the remote control device was as essential a tool of war as an assault rifle. Because there were so many aircraft from so many different nations participating in the daily air strikes, missions were parceled out to squadrons on a daily or even sometimes weekly basis, only further distancing the sense of a coalition being involved in a real shooting war.

A routine set in. Some air forces launched strikes after breakfast and returned before lunch. Sometimes the strikes were even carried out by remotely piloted aircraft, or drones, controlled from air bases in the continental United States.

But as removed and remote as the campaign was, the Inherent Resolve aerial campaign unleashed incessant and lethal strikes against the Islamic State. By October 1, 2014, not long after the alliance was created, the partner nations had flown 1,700 sorties against 322 Islamic State targets in Iraq and Syria. The strikes were largely aimed at what could be considered secondary targets: trucks, armored personnel carriers and machine gun emplacements.[74]

The air war grew in strength and scope throughout the autumn of 2014 and into the winter.

T. E. Lawrence was the first to discover the strategic splendor of the area around Azraq Castle in what's now northern Jordan, some one hundred miles north of Amman. The legendary Lawrence of Arabia realized the area's high ground had enormous value as a launching pad for biplanes to fly reconnaissance missions to assist the Allied push into Syria during the First World War. The Royal Jordanian Air Force opened a sprawling air base near the castle in 1981 and named it Muwaffaq Salti, in honor of a young Jordanian Hawker Hunter pilot shot down by an Israeli Mirage IIIC over the West Bank in 1966, a year before the Six-Day War. The base was one of Jordan's most important and certainly its most strategic. The northern perimeter fencing was a mere thirty-four miles from the lawlessness of the Syrian border—a distance traversed in seconds in a supersonic fighter-bomber. The Iraqi frontier was just one hundred miles away.

Muwaffaq Salti Air Base was one of the busiest air bases in the coalition against the Islamic State. F-16s from the US Air Force 20th Fighter Wing from Shaw Air Force Base, located some ninety miles northwest of Charleston, South Carolina, were based there, as were F-16s from the Royal Netherlands Air Force and the Royal Belgian Air Force;* F-16s from the Royal Bahraini Air Force operated out of the base, as well. There were, of course, other specialists who set up temporary headquarters at the base, including reconnaissance and satellite-imagery experts from America's military intelligence and security agencies, and, of course, men in civilian dress with pistols on their hips in the employ of the CIA. A special arrivals section, to complete all the paperwork required for entry into the country, was set up to handle the men and women arriving from the four corners of the world. It was the country's second busiest gateway after Queen Alia International Airport.

The different air forces were segregated at the base—each visiting contingent was assigned a separate area where the airmen and their support staffs had temporary housing, fueling and arming stores, and C3 (command, control and communications) facilities. The forces interacted in the operations center, where missions were given to the various contingents.

General Allen and his staff coordinated the multinational sorties with methodical efficiency. Air strikes from bases offshore, Turkey, Kuwait, Qatar and Jordan had to be timed on a tight schedule. There was no room for aberrations and technical glitches. The war against the Islamic State was a nonstop endeavor. The air strikes intensified throughout October and November. Multiple Islamic State targets were hit each day by coalition sorties. For the first two months of the campaign, the allied war was relatively clean and sterile from combat mission altitudes—especially from the perspective of hostile enemy aircraft and ground fire.

Each mission was assigned its own targeting cell consisting of base and squadron operations officers. They were respon-

---

* The Belgian armed forces called their air and ground mission against the Islamic State by the name Operation Desert Falcon.

sible to make sure that the pilots were mission ready and that all preparations, from a stock check of the ordnance to be carried to making sure that the intelligence was up-to-date and accurate, were in the sortie portfolio. Twelve hours before each strike, coalition forces involved in the operation would link up courtesy of secure video teleconferencing; participants included a representative from the CENTCOM hierarchy, a senior Inherent Resolve officer, and the various refueling and CSAR components assigned to the strike. Often the refueling squadron and the special operations force that would be summoned to rescue a downed pilot were in three different countries. Once the twelve-hour countdown began, the pilots were sequestered to their squadrons for rest until it was time for the final mission briefing.

Jordan was the only Arab nation in the coalition that performed kinetic missions against targets in both Syria *and* Iraq.[75] RJAF assets committed to Inherent Resolve were the F-16s in the No. 1 and No. 2 squadrons based in Muwaffaq Salti Air Base. An RJAF officer was permanently assigned to the Combined Air Operations Center, or CAOC, in al-Udeid, and he relayed the daily Air Tasking Orders, or ATO, to the Jordanian F-16 squadrons. The Air Tasking Orders consisted of the mission's call sign, the target, the suggested timing of the strike, air refueling coordinates, imagery, coordinates, the type of target and the intelligence that was available on the target.[76] The ATO also suggested the number of aircraft to be involved in the sortie.

The Jordanians flew their sorties in the morning. The Europeans flew midday. The Americans flew around the clock. The Jordanians usually attacked in pairs and were assigned targets from Sunday to Thursday—the days of a regular RJAF workweek.

The CAOC designated a weapons storage facility in Raqqa to the RJAF for the morning of December 24, 2014. It was Christmas Eve, and the base commander handed the strike to two pilots in the No. 1 Fighter-Bomber Squadron. First Lieutenant Moaz al-Kasasbeh, fate had it, was ordered to lead the mission.

Moaz al-Kasasbeh was born on May 29, 1988, and hailed from the tiny village of Aiy, located in the shadows of the Kerak Crusader castle, some fifty-five miles south of Amman. Little changed in Aiy in the years since the great Arab military commander Saladin tried to seize the Raynald of Châtillon's massive hilltop fortress in the battle for the holy land in the twelfth century. The village was carved alongside the twisting and turning hillside road; donkeys still roamed the main thoroughfare and shepherds still walked their flocks along the dusty embankments. A few dozen Bedouin families, all interconnected by blood and marriage, call Aiy home. The Kasasbeh clan, part of the influential Bararsheh tribe, was a prominent family in southern Jordan. Moaz's father was a retired professor; his uncle was a major general in the army. Moaz was one of eight children.

Moaz al-Kasasbeh enlisted into the RJAF on November 2, 2006, graduated from King Hussein Air College in 2009, and qualified to be a Viper driver, the affectionate universal nickname for a General Dynamics F-16 Fighting Falcon pilot, three years later. Narrowly framed and innocently handsome, Moaz had a naturally mischievous smile that concealed a very serious and religious young man. When dressed in a polo shirt and jeans, Moaz looked like a high school student. But his youthful appearance was deceptive: once his flight suit was worn and the Ray-Bans placed over his eyes, his face quickly revealed the swagger of a combat fighter pilot. He had dreamed of flying since he was a toddler. His simple village upbringing conflicted with the high-tech world of Mach 2 fighter-bombers and fly-by-wire controls. But he had been among the first in his class and a natural in the cockpit of the F-16. His flight instructor said that he had been the first aviator in a long time that had been ahead of his aircraft.[77] Lieutenant Kasasbeh was assigned to the No. 1 Squadron.

Unlike many of the veteran pilots in the squadron, Moaz lacked the swashbuckling arrogance that fighter pilots often possessed. His squadron mates recalled that he was serious, perhaps too serious for someone of his young age. "He wasn't like the rest of us," a squadron mate recalled. "He took everything with a serious yet simple stride. He was not a city boy, and he didn't speak with the cadence or the slang of someone who grew up in an urban center. Moaz was also shy and quite religious. He defined himself as someone who was guided by faith."[78] He refrained from coffee and off-color jokes. He did not smoke. In July 2014, Moaz married Anwar al-Tarawneh, a pretty engineer with piercing gray eyes. He talked in the squadron about the house he wanted to build for his wife and the family he planned. When he retired, he dreamed of becoming a farmer.

The newlyweds didn't have a lot of time together after their wedding—the RJAF went on high alert over the summer as the Islamic State consolidated its territorial gains in Iraq and Syria. Combat sorties against the Islamic State began in September. As Christmas approached, Lieutenant Kasasbeh was already a seasoned combat veteran, having flown five combat sorties against Islamic State targets in Raqqa and the town of Deir ez-Zor in eastern Syria. "He was thrilled to have been selected to spearhead the Christmas Eve raid," Brigadier General Ghassan,* the base commander, recalled. "The RJAF had not been at war since 1973, yet now they were the lead arm of a global coalition. Squadron morale was at an all-time high."[79]

The sun emerged over the hills to the east at precisely 6:33 a.m. on the morning of December 23. Lieutenants Kasasbeh and Salim had already said their dawn prayers, the *fajr*, at the first hint that the sun would appear to usher in a new day, and then headed for a quick breakfast of hummus, fava beans and a hard-boiled egg. The countdown had already begun. The two pilots were twenty-four hours from launch and there was a lot of work to do. There were

---

★   A pseudonym.

bureaucratic items to take care of: squadron paperwork to attend to, intelligence and weather jackets to be reviewed. The pilots conducted final checks on their personal gear, flights suits and survival equipment. They cleaned their Glock 19 9mm semiautomatic pistols that they carried inside holstered pouches, and they tested the batteries of their emergency transponders and radios. At noon they broke for lunch and then relaxed. Later in the day, to ease the tension and to temper the adrenaline, Moaz exercised on a stationary bike.

Darkness had blanketed the air base by 6:30. At launch-minus-twelve* the two pilots sat inside the ops center for the mass briefing. Airmen from several nations raced about, shuttling between rooms while juggling laptops, file folders and large thermal cups with piping hot coffee. Some of the US Air Force personnel wore Santa hats to usher in the holiday spirit; an artificial Christmas tree had been decorated with a heavy dose of glitter and Air Force–related ornaments. The two Jordanian pilots, senior squadron and base officers, as well as coalition liaison team leaders, sat in a small conference room where they were connected to similar teams of airmen in Turkey and Kuwait who would also be participating in the mission. The VTC technology was so good, the high-definition screens so sharp, that it felt as if all the components were sitting together in the same room.

An American air force intelligence officer reviewed the target and, more specifically, the latest collective intelligence concerning hostile forces in the area. Final drone and satellite imagery were analyzed to review threats, such as heavy machine gun or anti-aircraft missile or gun batteries positioned nearby; a careful last-minute analysis was conducted to make sure that there were no sudden movements of civilians near the targeted location. A coalition meteorologist reviewed the morning's weather, including expected winds and any last-minute storms that were expected.

Lieutenant Salim was Slider One-Two. The pilots reviewed their mission for everyone to hear and reiterated launch times,

---

* Twelve hours before takeoff.

refueling coordinates, target and the course they'd fly back to base. The refueling component of the strike belonged to the US Air Force KC-135 Stratotanker from Incirlik Air Base in Turkey. The ship commander reviewed the coordinates for the refueling linkup with his massive converted Boeing 707 that held as much as eighty-three thousand pounds of fuel for thirsty combat aircraft, including helicopters, that needed a midair top off to extend the range of their mission. A Stratotanker was always airborne to support one of the Inherent Resolve missions, hovering over somewhere safe at a high altitude; the slow-moving hulking tankers were always protected by fighter aircraft.

Perhaps the most important linkup by the VTC was between the two Slider pilots and their on-call CSAR standby force. A company of special operations US Marines who flew in MV-22 Ospreys caught the assignment. The Osprey came to the battlefield with a checkered history. The program to produce a multimission combat transport aircraft that combined the flexibility and functionality of a heavy-lift helicopter with the long-range and flight capabilities of a turboprop aircraft began in 1981. But the program was beset by fatal crashes, cost overruns and general questions as to if the Marine Corps, as well as other branches of service, needed an aircraft that was so expensive and so limiting—it could carry no more than thirty-two soldiers, and as was ultimately learned, it was too narrow to accommodate up-armored combat vehicles, even the small ones—in its constricting cargo hold. Still, urged on by generals from the Marine Corps, the Osprey program endured. The aircraft eventually entered service with the Marine Corps, the Air Force Special Operations Command and the Navy. A 2009 Government Accountability Office study reported that the Ospreys in Iraq had an abysmal full mission capable rating; the report also cited shown weakness in situational awareness, maintenance, shipboard operations and transport capability.[80]

The Slider target area in Raqqa was over seven hundred miles from Kuwait, however. Should one or both pilots be shot down or forced to eject, the airmen would have to evade terrorist forces for at least several desperate hours until the Marines in their Ospreys flew to the rescue; the aircraft would have to be refueled on the way to Syria from the Marines base in Kuwait.

The Marines assigned CSAR duties were to be supported by a flight of US Air Force A-10 Warthogs from the 122nd Fighter Wing, a unit from the Indiana Air National Guard, that would be on call in Turkey to provide close air support to any developing situation on the ground; the aging "Hog," a tank killer by design, was also brutally effective in grinding up anyone unlucky enough to be in its sights. Although it carried an arsenal full of ordnance, it was the aircraft's 30mm GAU-8/A Avenger autocannon and its depleted uranium armor-piercing shells that made it a battlefield game changer because it could simply chew up anything in its path: armor, fortification, human. The aircraft was economical, reliable, as strong as a tank and a lethal complement to the American war effort. Of course, it had been relegated to the Pentagon's chopping block on numerous occasions, though by the close of 2014, A-10s still were flying over 10 percent of coalition sorties against the Islamic State.[81]

The mass briefing lasted over an hour. Questions were tossed back and forth; officers jotted down coordinates onto pads while others inputted them into Panasonic Toughbook laptops. The RJAF contingent wished their coalition partners a "good hunt." The crews headed back to their squadrons for some more preparation and dinner.

Some of the ranking officers in No. 1 Squadron walked outside to smoke a last cigarette before lights-out. Nicotine was an accepted way to mitigate stress. Some of the pilots chain-smoked to excess. They gazed upward into the star-illuminated sky and checked all the boxes in the mental checklists they maintained before pilots flew into harm's way. Moaz said his nighttime prayers. He went to bed early that night barely able to contain his exuberance. Pilots wanted to fly, and Viper drivers wanted to be behind the controls of their fully armed aircraft hitting terror targets. Moaz sent an SMS home to Anwar and then turned out his light. The alarm was set for shortly after dawn.

The morning missions had become routine for No. 1 Squadron. The planning, preparation and execution were, after three months at war, committed to muscle memory. The process was a seamless exercise of muscle memory and efficiency. While the pilots still slept, groundcrews loaded AIM-9 Sidewinder heat-seeking short-range air-to-air missiles onto each of the two wingtips. The crewmen also placed five hundred rounds of ammunition for the M61A1 Vulcan cannon that was mounted inside the fuselage to the left of the cockpit; the M61A1 could fire six thousand rounds a minute and was ideal for strafing enemy emplacements. The Slider strike would drop eight 500-pound bombs—four for each aircraft—on the day's target.

Most of the coalition air forces struck Islamic State targets with precision-guided munitions. Known simply as "smart bombs," the ordnance was directed into the center of the crosshairs by laser, radio, satellite or infrared. Even though "dumb bombs" could be converted into the more precisely accurate variant with modification kits, the RJAF had trouble receiving the systems from the United States.

The pilots woke up and were showered before 5:00 a.m. Moaz attended to his morning prayers and sent an SMS message to Anwar. It was hard for the young pilot to take his mind away from his new bride back at home.

As the hint of daylight neared, the two men then proceeded toward the squadron headquarters for their preflight briefing. Lieutenant Colonel Ali* was the squadron's commanding officer. Short and thin with a proud military mustache and the swagger of a man who felt completely at ease inside an $18-million tool of war, Lieutenant Colonel Ali was the boss, chief tactician and

---

*   A pseudonym to protect his identity.

mother hen to a group of younger pilots, all serious men, but who felt that they were indestructible. Squadron commanders flew combat sorties with all their pilots; Lieutenant Colonel Ali knew the capabilities of each of his fliers. He trusted each one to lead a strike or to serve as his wingman, but he worried. These men were *his* men. Often he worked throughout the night before an operation reviewing every detail of the mission.

The briefing commenced at 5:30. Lieutenant Colonel Ali ran it like a stern professor. He addressed the two Slider pilots, along with the group's intelligence and operations officers as well as some of the other pilots from the group. He spoke from behind a podium positioned underneath framed color portraits of King Abdullah and his father, King Hussein. The squadron flag decorated one of the walls.

The briefing amounted to a quick summary of the mission's technical landmarks and was more of a reiteration than an explanation. Moaz and Salim paid close attention to the technical bullet points: altitudes, coordinates, latest weather updates and the last-minute intelligence updates. The squadron commander ended the briefing with a rousing "good luck." The pilots then suited up. They put on their G-Suits, survival harnesses and flight bags. They double-checked their sidearms and their survival knives. It was forbidden for the pilots to take any personal items with them on the mission other than their military ID cards. Moaz's number was 52166; his blood type was A-positive.

Before the quick bus ride to their armed, fueled and ready aircraft, Moaz wanted to send Anwar one final message. He did it discreetly so that the others in the squadron wouldn't see him. It was a terrible thing to be young and in love in a combat formation, and he wanted to avoid the good-natured but biting commentary. But there wasn't enough time to return the phone to his locker, so he slid it into his flight suit. It wasn't a big deal, he thought to himself. After all, he was supposed to be back before lunch.

A white minibus drove the pilots for the quick trip to their

awaiting F-16s. Each pilot conducted a walkaround of his aircraft. A small metal ladder was already affixed to the left side
of the aircraft that the pilots had to climb up, and with the help
of a ground crewman, then squeeze and slide their way into the
narrow confines of the cockpit. Once they were strapped in, the
pilots put on their flight helmets. The pilots shook the hand of
their crew chief, who lowered himself down and then moved
the ladder away from the aircraft. Moaz checked his controls
and then lowered the reinforced glass canopy down to create a
tight seal over the cockpit.

The two F-16s had once belonged to the Belgian Air Force.
Both aircraft taxied to the weapons section where their missiles and bombs were armed. Once their weapons systems were
hot, both aircraft moved slowly toward the takeoff line. The
sun was slowly emerging and painted an orange glow on the
darkened tarmac.

"Slider One-One requesting departure," Lieutenant Moaz
radioed the tower. Slider One-Two was several meters behind
him and to the right. There were RJAF and coalition personnel in the control tower monitoring computer screens, the radar
and other data inputs. "Slider One-One," the air traffic control
officer replied, "departure confirmed." Lieutenant Moaz fired
up his afterburners as did Lieutenant Salim. Each F-16 boasted
Pratt & Whitney F100-PW-200 afterburning turbofan engine
with 23,830 pounds of thrust power, and the roar each produced
on takeoff was deafening.

One hundred years earlier, when Lawrence of Arabia launched
reconnaissance aircraft into Syria from the same spot, the bi-
plane could reach a maximum speed of seventy-two miles per
hour. Now the F-16 quickly disappeared in a flash and an ear-
splitting roar straight into the morning sky.

The Slider aircraft headed east and then north, crossing the
border beneath them into Iraq. The two F-16s followed a cir-

cuitous course over territory that the government in Baghdad owned and then rendezvoused with the US Air Force KC-135 tanker. All sorties into Syria flew via the Iraqi corridor. It was considered the safest, especially since Hezbollah units in the south and west were amply supplied with advanced antiaircraft weapons, including shoulder-fired missiles, and Russian pilots, flying covertly on behalf of the Assad regime, protected the skies over Syria's major cities.[82]

It took thirty minutes for each aircraft to take on five thousand pounds of fuel in midair—the top off was just enough to allow the F-16s to bank west and head to Raqqa, drop their bombs over the target and then push southwest flying a predetermined zigzag course back to base. The weather was perfect for a morning strike. The ride was smooth and uneventful.

The two F-16s reached the outskirts of Raqqa at approximately 9:00. The course they took from the refueling rendezvous was circuitous and classified. The flight path was blueprinted to limit any semblance of routine between this mission and the ones that preceded it. The aircraft approached the city from the north and then banked a hard right, then a hard left to come in and attack from the east. The city stretched as far as the eye could see from four thousand feet ASL, or above sea level. Slider One-One and Slider One-Two readied themselves for the bombing run.

At one time, a quarter of a million people called Raqqa home. The city, on the northern banks of the Euphrates River in northern Syria, was now the capital of al-Baghdadi's Caliphate. Raqqa, once an Ottoman customs outpost that connected Iraq, Syria and Turkey through trading routes, had for centuries attracted the best and the brightest as the star of Constantinople's eastern provinces. The Kurds established a foothold in the city centuries ago; the Circassians came to Raqqa fleeing persecution at the hands of the Russian czar; and the Armenians came from the starvation and slaughter during the forced marches of

1915–17 in the Ottoman perpetrated genocide. Raqqa was a city of commerce and spices, and of cash deals and coexistence. The city survived the rule of their French colonial masters, the fighting of the Second World War, and numerous coups and political crackdowns by Hafez al-Assad, Bashar's father and a man who dealt with his enemies by razing their cities and killing everyone inside.

The al-Nusra Front, along with other rival militias, seized control of Raqqa in the summer of 2013, evicting pro-Assad forces from the city and the outlying areas. Sharia courts were established in the city's football arena for the execution of Alawites, Shiite volunteers from Lebanon and other enemies of the Caliphate. Islamic State forces soon became the dominant armed entity in the city, controlling a constant flow of oil and commerce along the same routes that traders had used for centuries to connect Syria with Asia, the Middle East and Europe.

Satellite and aerial reconnaissance imagery of Raqqa, essential tools that Inherent Resolve commanders needed in order to assign targets to the air contingents, provided accurate information on where the Islamic State hid its weapons, stored its ammunition and housed its fighters; the uncanny detail of the high-resolution PHOTINT made it possible for intelligence officers in Kuwait to read the brand of mineral water carried in the back of the white Toyota Land Cruisers—the chariot of the Caliphate's mobile and merciless army—that raced up and down the streets of the city with a heavy machine gun welded into the cargo bay. Of course, the other intangibles of this intelligence-gathering effort were left to the spies. They were the ones who could detail just how many of the city's residents were estimated to have fled. They knew where the taxi drivers went to eat and where the foreign fighters from France and Tunisia hung out. The spies were the ones who knew that people were praying in the streets not only because the mosques were full but because Islamic State enforcers, some carrying medieval truncheons and maces, beat to death merchants who refused to close their shops every time the call to prayer was heard.

Every attempt possible was made to make sure that coalition strikes hit military targets only, but there was always collateral damage. Buildings that were not intended to get hit suffered damage. Civilians not known to be near a target were within the blast radius of a strike. Innocent people died.

The Slider sortie targeted a warehouse-type building on the eastern side of the city that was used as an Islamic State arsenal. The building, the intelligence reported, housed tons of explosives and ammunition, along with rifles, heavy machine guns and even pieces of towed artillery.

Slider One-One radioed the position of the tandem several minutes before engagement. The pilots flew in tight combat formation as the target quickly neared. Enemy activity on the ground was not reported. The two Jordanians' aircraft saw the target in their sights. Lieutenant Moaz led the stick, maneuvering his dive and then unloading the four MK-82 unguided low-drag general-purpose bombs. Lieutenant Salim followed close behind. The F-16 was suddenly lighter and more agile once the ton of bombs was no longer carried by the aircraft's small but powerful frame. Both aircraft pulled up quickly and then leveled off at the prescribed altitude before heading back. The Islamic State weapon storage facility was obliterated by the bombs dropped by the two Jordanian pilots. The bombs hit close to where intended. The explosives and material inside the warehouse evaporated into a fireball and then a blinding cloud of acrid black smoke that covered the sun.

Neither defensive system on board the aircraft indicated that an antiaircraft missile had locked onto their Vipers. The pilots didn't see the telltale white-hot blips of antiaircraft fire being sprayed from the ground to the heavens. It had been smooth sailing—an average mission, until, as the aircraft placed Raqqa behind them, a fire erupted in the engine of Slider One-One.

Efforts to extinguish the fire failed. Lieutenant Moaz began to lose altitude. He began to lose control of the aircraft. Systems quickly began to fail.

Pilots will always try to make it as far from harm's way as they can when their aircraft have been hit or when they suffer a catastrophic malfunction. Moaz was experienced enough in the F-16 to know that he didn't have a prayer to make it back to base; his controls were spinning and unresponsive and his throttle failed to respond to his efforts. There also wasn't anywhere nearby where he could ditch his aircraft close to friendly forces. Virtually all the forces in Syria were hostile to the coalition. The Islamic State had placed a hefty bounty on the head of an enemy pilot.

Moaz radioed in a Mayday signal that was relayed to the control tower at Muwaffaq Salti Air Base. Word was received in Kuwait by the coalition ops officer. Moaz was losing altitude at a rapid rate. He had tried to stabilize the aircraft, but the loss of engine power proved to be catastrophic. Pilots lived by the philosophy that if they could not determine the attitude or capability of their aircraft, then they must eject. Pilots are continually taught that if there's any doubt about the status of their aircraft, they should leave it. The thought process of the "what ifs" doesn't happen in the air; they occur on the ground before each sortie.[83]

There was no panic on Moaz's voice over his direct frequency with Lieutenant Salim. He yelled "Bailout, bailout, bailout" into his oxygen mask and then unfastened it from his helmet. He reached for the ejection handle and braced himself for being flung out of the aircraft in a manner that some pilots have compared to being fired out of a shotgun. The aircraft's canopy was jettisoned; rocket boosters lifted the flight seat and pilot straight up in a violent burst of energy. The winds at 350 knots slapped Moaz's face with thunderous force; at four thousand feet ASL, Moaz had precious little time before the chute opened and he dropped to earth. There were many things that could go wrong with the olive nylon canopy—the chute lines could twist, or they could catch to the damaged aircraft and bring the pilot to earth in a crashing explosion. Moaz drifted quickly to earth and he attempted to direct his chute toward the banks of the Euphrates. Pilots train extensively for low-altitude ejections because they tend to hit the ground hard. The proper position requires the pilot to aim his body to one side and then collapse on the fall to dissipate the pressure of the impact. But the wind directed him toward the water near the village of al-Ekeersha. He tried to guide his chute but was unsuccessful in the harsh winter wind, and he splashed into the frigid water. Moaz hit the cold river water hard. He unfastened himself from his harness, but his body was badly bruised in the ejection; it is likely that he tore ligaments and suffered a fracture or two. He flopped uncomfortably as he hit the water; he splashed hard. Pilots are trained to recover from a landing—even a harsh one—and he tried to grab his radio transmitter and his Glock. He glanced toward the skies, searching for Lieutenant Salim, who was flying close cover overhead. The roar of Slider One-Two's F-16 provided a momentary sense of protection.

The US Marines in Kuwait reportedly were never even scrambled into action. Moaz was in Islamic State custody by the time Ospreys would have been airborne.

A force of Islamic State fighters had seen the RJAF in distress in the skies above Raqqa and they raced toward the river where the parachute fell to earth. Some of the fighters, wearing desert tan camouflage fatigues, aimed their AK-47 assault rifles and joked to see who would pick off the pilot, but Abu Bilal al-Tunisi ordered them to shut up and put their weapons on safe. Al-Tunisi was a senior Islamic State military officer in Raqqa known for a sadistic smile and a barbaric lust for watching others suffer. Al-Tunisi, as per his eponym, hailed from Tunisia, and like many foreign volunteers from that North African state, he showed a penchant for violence. He was a short squat man with a chubby face and a wild, bushy beard. He wore a gray, black and white beanie and American-style military BDU (Battle Dress Uniform) camouflage fatigues, stolen from Syrian Army stores the Islamist army had overrun. He looked more like an out-of-shape weekend paintballer than a Caliphate field commander, but he knew the value of a live prisoner.

Lieutenant Moaz al-Kasasbeh had no chance to evade capture, or to resist. By the time he gathered his wits to try to call in his position, six men with AK-47s were already in the water with him with their weapons aimed at his head. Al-Tunisi's men beat Moaz as he was dragged out of the water to shore. The gunmen ripped off Moaz's flight suit, stripping him of everything but his white undershirt in a humiliating display. Al-Tunisi's bodyguards rushed into the ice-cold water to hasten the capture; the men gazed skyward, their weapons at the ready, expecting to see helicopters with rescue commandos flying fast and low toward them from over the horizon. Al-Tunisi took the time to remove his combat boots, surplus jungle shoes seemingly bought online, so that he could venture into the water and be seen dragging out a disoriented and frightened Moaz. The pilot's face bore the bruising of ejection and the butt end of several rifle blows. His lips were swollen and bleeding.

Al-Tunisi smiled in gleeful glory realizing the importance of the moment. Islamic State fighters whipped out their Chinese-made Huawei smartphones to photograph their prisoner.

BOOK TWO

THE HUMAN BAZAAR

5

Held Captive

War is wretched beyond description, and only a fool or a fraud could sentimentalize its cruel reality.

—Senator John McCain[84]

THE CONTROL TOWER was busier than usual that morning at the Muwaffaq Salti Air Base. Nervous officers watched with field glasses as the F-16 made its final approach. The room was crowded. There were men in US Air Force camouflage who were standing close to their Jordanian counterparts. They talked on satellite phones to inform the next higher up in the coalition command—in Kuwait and at CENTCOM headquarters half a world away in MacDill Air Force Base, in Tampa—that the one pilot was near. There were also men in civilian dress with laminated IDs around their necks. They, too, were on their smartphones, sending messages to their headquarters wherever they might be. Sorties went up in pairs and they returned the same way.

Lieutenant Colonel Ali was trying to kick his nicotine habit, but very few men ever managed to quit smoking during wartime. There wasn't a thing that happened inside No. 1 Squadron that wasn't his worry and responsibility. It was a daunting task, and one he did not take lightly. The squadron commander oversaw a dozen or so alpha males who felt that they were Mach 2 supermen; Viper drivers were blessed with an overabundance of confidence that outsiders could easily mistake for arrogance. But the braggadocio masked vulnerability, and in a combat environment that meant concern. When he wasn't flying combat sorties himself, Lieutenant Colonel Ali listened to the radio throughout each squadron mission to monitor his pilots. His was the largest office in squadron headquarters. The mood inside was grim.

The Islamic State's propaganda machine tweeted images of Moaz's capture thirty-five minutes after he was pulled out of the frigid waters. The shock felt in Jordan was numbing. Lieutenant Colonel Ali remembered feeling that someone had punched him in the gut.[85]

Every member of the squadron felt paralyzed by uncertainty once word spread that Moaz had been forced down over enemy territory. The scene was eerily reminiscent of RAF pilots checking the skies to see if their mates were returning from duels with Luftwaffe fighters and bombers over London during the Battle of Britain. The pilots wanted to know what happened.

Slider One-Two touched down alone just before 11:00 a.m.

A convoy of staff cars, Suburbans and Pajeros awaited Lieutenant Salim's wheels down. Base commander Brigadier General Ghassan stood in silence. So, too, did the coalition operations and intelligence officers assigned to the Jordanian base who were alerted about the loss of an F-16 over Raqqa. Salim taxied the F-16 to its shelter, where an aircrew was waiting. The young pilot unbuckled himself from the constricting confines of the cockpit, climbed down the stairs. The generals and colonels looked impatient. Salim removed his helmet and saluted.

He shielded his eyes from the winter sun that blazed overhead, providing a semblance of warmth on a cold day.

Brigadier Ghassan ushered Salim into his staff car. The young pilot was rushed back to the base commander's office near the main gate. Some of the senior pilots from No. 1 Squadron were already waiting there in their flight suits. They wanted a mission, any mission, to fly cover over the crash zone where Salim last saw Moaz and help to facilitate a rescue. There was anger expressed as the debriefing commenced. Where was the rescue force? Why wasn't anything done the moment Moaz informed flight control that he needed to eject? There was an effort to restore calm, but the room had erupted into shouts in both Arabic and English. Armed sentries posted in front of the front door heard the commotion inside.

A call was made to Marka Air Base where Major General Mansour al-Jbour, the RJAF commander, was sitting with members of his operational staff. Calls were also made to King Abdullah's military staff. The monarch would need to be notified at once.

King Abdullah was celebrating Christmas Eve with the Christian community in Kerak that day. A festive lunch had been prepared for the royal entourage and the town's religious leaders; a children's choir sang festive carols in angelic harmonies. The celebration had been on the calendar for months and was an annual event. A military aide in camouflage fatigues walked toward the dais and whispered in King Abdullah's ear that an aircraft had been lost over Syria. The king left the celebration early. A helicopter was summoned for the short flight to JAF headquarters on the outskirts of Amman. Lieutenant General Mashal al-Zaben, the chief of staff, was waiting at the helipad.

A roundtable of air force, army and intelligence personnel briefed the king. The details of the flight, the mission and the coalition resources were laid out in sharp bullet points. The immediate question of why a CSAR effort was never launched was an explosive talking point. The king looked around the room and asked the most pressing question: "How do you think this is going to end?" The prediction in the room was that in a few days Moaz would be featured in a YouTube video wearing a jumpsuit, making a confession before being beheaded.[86]

The king ordered his military and intelligence chiefs to provide him with options. He also wanted the GID to identify anyone who touched a hair on Moaz's head. If the pilot was not going to be returned safe and sound, there would be retribution.

Safi Kasasbeh, the family patriarch, and the other male heads of the clan were waiting in an adjacent room, brought to headquarters by an air force liaison officer. The family was notified the moment the Twitter feed of Moaz's capture was verified; the men were informed by the Royal Court, and then transported to general headquarters to meet with air force personnel and, ultimately, the king. Some of Moaz's family were armed forces veterans; the others knew how the Middle East worked. Safi Kasasbeh had no illusions concerning what his son's fate would be. When he and the other men met with King Abdullah behind closed doors, the men spoke frankly. "Moaz's father told me, 'Do what has to be done and make them pay.'"[87]

Alice Wells, the fifty-one-year-old US ambassador to Jordan, tried unsuccessfully to phone the Jordanian monarch the moment she received word that an aircraft had been lost over Raqqa. She was the daughter of an American diplomat; born in Beirut in fact, but her Arabic was wobbly even though she had been posted to the US Embassy in Riyadh years before and had spent a career in Foggy Bottom's Near East Division. Wells was described by some on the Amman cocktail circuit, including veteran journalists who had covered the region for a quarter century, as a brave diplomat.[88] But in the eyes of the Jordanian leadership she personified the direction that American policy in the Middle East was taking in shifting away from the Arab world.[89]

The first reports of what happened over Raqqa reached the Beltway before dawn. It was the day before Christmas and many of the various branches of government that needed to be manned twenty-four hours a day were operating with a skeleton crew. With Christmas falling on a Thursday, it was easy for the higher grades to turn the holiday into a five-day vacation. Some with coveted G-13 and above pay wouldn't be back before the New Year.

News that an F-16 and a Jordanian airman had been lost over Raqqa reached Langley at around 5:00 a.m. on the JWICS (pronounced JAY-wicks) network for top secret communications.[*] The emails to the Pentagon and the diplomats at the State Department were sent on the less secure Secret Internet Protocol Router Network known as SIPRNet; years earlier, a veteran CIA hand commented, such news would have been transmitted via encrypted FLASH cables noting their urgency.

A White House national security staffer traveling with the president was informed, as well. It was just after midnight when the top secret email traffic began to crisscross time zones. It was midnight in Kailua Bay, Hawaii. The Obama family had left Washington, DC, on December 19 for two weeks of well-deserved respite highlighted by golf, surf and sand, and bowling. News of the incident was held until Obama woke that morning. The president huddled with the holiday crew that accompanied him on vacation. President Obama phoned King Abdullah later that morning pledging American support for Jordan and asking for restraint and perseverance. It was essential, the Americans urged, that the coalition's cautious pace not be interrupted. The United States, Obama pledged to the Jordanian monarch, would do everything in its power to help secure the pilot's release.[90]

The American president understood that the Islamic State

---

[*]   The State Department and the Department of Defense used SIPRNet, or Secret Internet Protocol Router Network, the interconnected computer network used to transmit classified information, up to and including items marked "Secret."

would amplify the importance of the captured Jordanian pilot and turn his plight into a major military and political propaganda weapon. Arab participation in the coalition was essential to denying al-Baghdadi and his jihad legion any chance of dictating the narrative in a splintered and tenuous Arab world. It was imperative that one pilot's fate not be allowed to muddy the waters of what was so far a successful international effort against terror.

The Islamic State gunmen who dragged First Lieutenant Moaz al-Kasasbeh out of the riverbed examined his flight suit once he was out of the water. Abu Bilal al-Tunisi's hunter squad demanded that Kasasbeh strip down in the water to show that he had no hidden weapons on him; pulling a half-naked and terrified young man out of bone-chilling water also made for terrific visuals. They found his wallet, his military ID card and the smartphone he forgot to leave behind. The gunmen made a point of photographing the event with their phones. They laughed and joked as Moaz was kicked and beaten. Some of the men spit on him and cursed his mother in various dialects of Arabic. They aimed their weapons at him as he shivered from the cold water. He was on his knees as he stared at the men who surrounded him, bewildered and squinting into the winter sun. There wasn't time to ask for mercy. He was thrown into the hold of an American-built MRAP armored troop carrier, one of the spoils of war seized in Mosul, and then driven from the crash site to a secret location somewhere in Raqqa.

Abu Bilal al-Tunisi was giddy when he radioed the command staff in Raqqa that he was returning to the city with a trophy. He laughed when some of the fighters, including a young Tunisian jihadist believed to be Hamza Maghraoui, stood on Moaz's head, grinding the hardened rubber soles of his black combat boots into his scalp. Maghraoui was part of the Tunisian legion who were once part of al-Qaeda in the Maghreb and then, looking for a war, joined the al-Nusra Front before signing up with the Islamic State.[91] Moaz was kicked and taunted inside the belly of the MRAP. The thick North African accents were hard for the terrified Jordanian pilot to understand. The Tunisians made a point to blindfold him before reaching their destination.

Moaz was taken to one of the former government buildings in Raqqa now occupied by religious elders. He was blindfolded and spit on as he passed the legion of fighters who met his arrival. He was punched and kicked. Some of the fighters laughed and uttered that a dead man was walking by them. Abu Bilal al-Tunisi held his prey tightly, determined not to let any of the other fighters or regional commanders put their hands on Moaz and try to take credit for the capture. The Tunisian was known for his eagerness to show battlefield prowess. The Jordanian F-16 pilot would make him a legend. He had already forwarded the photos from the phones of his men to the propaganda arm of the organization where, shortly, it would be tweeted out to the entire world. He would soon be famous.

Al-Tunisi brought his prisoner to the basement where men were held and where Abu Luqman al-Shweikh, the most powerful local commander in Raqqa, was waiting.

Abu Luqman's real name was Hawikh al-Moussa Ali,[92] a Syrian from the prominent Ajeel clan of Raqqa, who had been released from one of Syria's torture academies during a general amnesty meant as a goodwill gesture following the eruption of the Arab Spring. He had been brutalized by the secret police in the notorious Sednaya Prison. The experience left him scarred and psychopathic. He was known to have a hair-trigger temper. Abu Luqman was the wali, or governor. Nothing transpired inside Raqqa without his knowledge, and as a true mafia warlord, he earned tribute from the spoils of war that passed through his city. He earned top dollar for every barrel of oil sold from the refineries under Caliphate control. Money made him even more powerful than some of the other men who ran the war.

Not all the Islamic State's prisoners were inspected by members of the Caliphate's hierarchy. Most captives were simply beaten and robbed, then taken to a ditch somewhere and murdered. The high-profile hostages, non-Arabs seized in Iraq and Syria, were beaten and tortured and held for special treatment. The first Western prisoner executed was James Foley, an American freelance journalist who was seized in Syria in 2012 and ultimately landed in the hands of the Islamic State. Foley was tortured, forced to convert to Islam, forced to wear an orange jumpsuit like the al-Qaeda terrorists held at Guantanamo Bay in Cuba[93] and ultimately beheaded by an infamous British fighter nicknamed Jihadi John, a Kuwaiti-born executioner of Iraqi descent who had been raised in Queens Park in London. Often in the videos of the beheadings, the executioner rambled on with some political statement. When, in September 2014, American-Israeli journalist Steven Joel Sotloff was murdered, Jihadi John said the following on camera: "I'm back, Obama, and I'm back because of your arrogant foreign policy towards the Islamic State."[94]

David Haines and Alan Henning, two British aid workers, were beheaded days later in the same ghoulish manner. All the non-Arab prisoners were also subjected to intense psychological torture. According to one French intelligence officer, a former Foreign Legionnaire who worked the area close to the front lines of the civil war, "The abuse was done for pure sport and entertainment. The cruelty displayed to these poor and condemned men was beyond measure. The bastards only tortured those they intended to murder. The merciful thing to do with these doomed men was to simply put a bullet in the back of their heads."[95] Al-Shweikh wasn't one known to display mercy—especially when it came to a high-profile prisoner. He was the one, coalition intelligence believes, who ordered the beheading of Haines and Henning.

Moaz was thrown onto a chair with his hands secured behind

his back with plastic flexicuffs in an uncomfortable full nelson position. He was asked the routine questions asked of all prisoners of war: name, age, rank, base and next of kin. The questions were superfluous, of course. Moaz's ID card had the information on it. The questioning continued, though, as did the beatings. It didn't matter to the governor that Moaz, like him, was a son of the desert, a cherished son of a prominent clan. To the Islamic State, Moaz was worse than the Shiites and the Crusaders who came to fight and die in Iraq and Syria, and they had the means to turn that treachery into lethal propaganda. One of the main objectives of the Islamic State, beyond the expansion of the Caliphate and the vanquishing of its enemies, was to control the narrative and reach a global audience.[96]

Besides being the Raqqa governor, Abu Luqman al-Shweikh was responsible for all the foreign fighters in the city. They came from all over the Arab world and from all over the Sunni world to kill and be martyred. A Tower of Babel worth of languages could be heard on the streets of Raqqa: the armed mem spoke Arabic and Turkish, French and Chechen, Albanian, German and English. Many of the foreign fighters had advanced computer skills, and these men joined former Iraqi and North African counterintelligence officers who now swore allegiance to Abu Bakr al-Baghdadi to head the Islamic State's IT Division. Al-Shweikh ordered superfast and powerful computers and had them smuggled through Turkey or traded for oil. The computer labs ran 24/7 in Raqqa, Aleppo and Homs. The jihad techies were engaged in all the technological needs any criminal enterprise would require. Al-Shweikh's men were also top-flight hackers. It didn't take long for Moaz to cough up the pass code for his Facebook accounts after being savagely beaten.

The techies uncovered a treasure trove of incriminating and invaluable intelligence from Moaz's Facebook and Instagram accounts, and a wealth of personal photos, including some taken at the air base with him and his squadron mates. The images were, of course, private. They were the memories of warriors at work and at play and never meant to be seen by anyone outside the immediate brotherhood. The data on the Royal Jordanian Air Force constituted a serious breach of operational security, as did any information it revealed about activities against the Islamic State launched by the coalition.

Moaz's primary value, though, was to the Islamic State's weaponized propaganda. Moaz was to be used to disrupt the Arab participation in the coalition and to rally Jordanians into its ranks. The Islamic State's mass media machine produced films, web content and audio CDs using the latest technologies. Cameras, sound and lighting equipment, as well as editing machines and Apple laptops and desktops were smuggled through Turkey. So, too, were the latest models of smartphones (primarily Chinese-made Xiaomis and Huaweis). The media arm made serious use of the encrypted Russian-made app called Telegram to disseminate videos and pictures, as well as operational instructions. A social media unit disseminated messages and videos on Facebook and Twitter.

Cyberspace was a battlefield in which the Islamic State excelled. Internet channels, especially those that had the capacity to show videos, were critical to spreading the word. Islamic State fighters on the front line were often issued with GoPro cameras they could affix to their helmets or their assault rifles so that rousing combat imagery could be scored with music and uploaded online. The propaganda front was a critical arm of the Caliphate's appeal. Posters that looked like they were designed by Madison Avenue were plastered throughout territory captured by jihadi forces and proclaimed, "O media man, you are indeed a jihad fighter. The media campaign is no less important than the battle being waged on the battlefield. Each of you [media professionals] must be vigilant and take every opportunity to renew the intention [to act on behalf] of the Islamic State."[97]

The Caliphate's media operation was likened to CNN and

Britain's BBC—breaking news happened 24/7. The budget for the office in Raqqa alone topped $2 million a year.[98]

The Islamic State's official spokesman was Taha Sobhi Falaha, who was better known to the intelligence services by his nom de guerre of Abu Mohammed al-Adnani. A native of Idlib in Syria and a veteran of Assad's notorious prisons, al-Adnani ventured to Iraq in the wake of the American invasion to fight with Abu Musab al-Zarqawi's variant of al-Qaeda in Iraq. Known as a capable field commander with a brilliant commando mind-set, his exploits in the Fallujah were legendary. Like many in the Islamic State's hierarchy, he was captured in the American-led counterinsurgency surge of 2005 and detained for five years in Camp X-Ray. Al-Adnani wore many hats inside the organization's leadership. In addition to serving as the Caliphate's mouthpiece and chief copywriter, he also headed its special operations and was the chief of all external operations, controlling the sleeper cells and foreign fighters in place around the world. Al-Adnani was the mastermind behind the beheading videos released with glee to the major media outlets. In November 2014, footage of al-Adnani's men beheading twenty-two Syrian soldiers seized in the civil war was tweeted out to the world.

Now, Al-Adnani viewed all the images taken earlier that day when Moaz was captured. He ordered the techies to send the photos out on Facebook and on Twitter. The world now knew what King Abdullah, President Obama and General Austin had been grappling with all morning and afternoon.

News of the captured pilot's identity spread fast, and reporters—local and from the international press—rushed to sleepy Aiy, kicking up a dust storm of breaking news in the hope of getting a statement. Safi Kasasbeh greeted reporters in his living room wearing a traditional brown Bedouin galabia. "I do not want to think of my son as a hostage," he told a small army of reporters that had made the pilgrimage. "I call him a guest. He is a guest among brothers of ours in Syria's Islamic State. I ask them in the name of God and I ask them with the dignity of the Prophet Mohammed, peace be upon him, to receive him as a guest and to treat him well."[99]

Later that evening on the evening news, Jordanian state television said that the government warned the terrorists not to harm the captured pilot. CENTCOM commander General Austin said, "We will support efforts to ensure his safe recovery and will not tolerate ISIL's attempt to misrepresent or exploit this unfortunate aircraft crash for their own purposes."[100]

The Islamic State produced an online magazine called *Dabiq*, which was published in Arabic, English, German and French. The magazine's primary editorial message was to rally the believers against Jews, Christians, Hindus and Shiites. The first issue was released in July 2014. The online publication was named after a small town in northern Syria where, according to some politicized and questionable interpretations of a hadith attributed to the Prophet Mohammed, an apocalyptical battle would be fought between victorious Islamic armies and what were termed "forces from Rome."[101] The end of days subtext was one of the magazine's main appeals, as was its striking photography and rousing articles that helped recruit foreign fighters from the four corners of the world.

Lieutenant Moaz al-Kasasbeh had been a prisoner of the Islamic State for less than a week when the correspondents from *Dabiq* interviewed him for the cover story of their year-end issue. The publication was released on December 30.

In the brief yet telling article, Moaz is referred to as al-Kasasbeh the murtadd, Arabic for an apostate. The article showed a photo of Moaz wearing orange inmate pajamas. His face was swollen; GID specialists and CIA officers in Amman who reviewed the online article were certain that he had been subjected to harsh and regular beatings. "My plane was struck by a heat-seeking missile," Moaz confessed, sticking to the Islamic State narrative that their gunners had blown the F-16 out of the sky. "The system display indicated that the engine was damaged and burning. The plane began to deviate from its normal flight path, so I ejected."[102] He also reflected on life at the air base with American servicemen and sharing meals with the US coalition partner members. "The Americans sometimes have dinner with us and eat Mansaf* which they liked a lot. Their talk does not include details about operations because of matters of secrecy and security."[103]

When asked if he knew what the Islamic State would do with him, the response was telling. "Yes... They will kill me."[104] After the article posted online, Islamic State accounts on Twitter asked followers to suggest the most desired way for Moaz to be executed. Some suggested beheading. Others called for a firing squad. One person even suggested that he be steamrolled. A hashtag that translated as "We all want to slaughter Moaz" was tweeted tens of thousands of times.

---

* The national dish of Jordan consisting of lamb cooked in a sauce of fermented dried yogurt and served with rice.

6

The Point of No Return

The spark has been lit here in Iraq, and its heat will continue to intensify—by Allah's permission—until it burns the crusader armies in Dabiq.

—Abu Musab al-Zarqawi[105]

THERE HAD TO be a military option, King Abdullah calculated, some sort of spectacular rescue operation, that could be mounted to bring Moaz home alive. The Jordanian monarch was a soldier at heart—one who was disheartened to learn that as monarch he wouldn't be allowed to parachute, both static line and freefall, with his men. Although he began his military career as a tank officer, the special forces were his passion, and he served in a commando battalion and then as commando brigade commander. Promoted to the rank of general, he put the pieces in place to create RJSOCOM—the Jordanian version of the US Special Operations Command based at MacDill Air Force base in Tampa, Florida. As a colonel and later a flag officer, Abdullah saw to it that Jordan's tip of the spear had the best training and equipment possible. He personally enhanced the capabilities of CTB-71, the army's counterterrorism and hostage-rescue unit. The king saw to it that "71" trained and interacted with the top teams in the world, including the US Army 1st Special Forces Operational Detachment-Delta,* US Navy DEVGRU,† Britain's 22 Special Air Service, France's GIGN and Canada's JTF-2. Since the establishment of KASOTC, a world-class special operations training center near Amman in 2009, Jordan had become a hub for the world's top commando and counterterrorist units to share tactics and innovations.

Commando operations were always equations of risk versus reward with the ever-present asterisk of Murphy's Law. The commanders of the special operations forces had to plan and prepare missions that could be articulated in a presentation, as well as be tactically possible in practice. The blueprint of daring was always perfected on models. Rehearsals of the real thing transformed theory into live-fire practice. All the pieces of the operation were brought together in a dress rehearsal. Most important, they had to sell the plan to the decision makers. Politicians and generals had to authorize a strike with so many precious assets— men and machines—that would have to cut a path deep behind enemy lines in order to carry out an operation that was fraught with risk. Their thought process consisted of harsh calculations of strategic necessity and political fallout. Operations that are successful can rally a nation and become national legends: Entebbe, Prince's Gate and Abbottabad have become synonymous with boldness, imagination and unprecedented daring.

The special operations side of the Inherent Resolve coalition was ready to go into Raqqa and extricate Moaz. The coalition had some of the world's best special operations units on call in

---

* Also known as the Delta Force, the Combat Application Group, and Task Force Green.

† The legendary SEAL Team SIX maritime counterterrorist unit that carried out the bin Laden raid in May 2011.

the anti–Islamic State alliance, and they never shied away from a mission.

Months earlier President Obama gave the okay for two units inside the Joint Special Operations Command, or JSOC, to enter Syria to try to rescue James Foley and other American hostages held by the Islamic State. Twenty-four Delta operators were flown to northwestern Syria by the famed Night Stalkers of the army's 160th Special Operations Aviation Regiment (Airborne).[106] The operators ventured deep into enemy territory in both rotary and fixed-wing aircraft; surveillance aircraft flew overhead. The White House never had any intention of revealing mission details, even though many Islamic State gunmen were killed or wounded in the exchanges of gunfire. "We never intended to disclose this operation," National Security Council spokeswoman Caitlin Hayden revealed in a statement. "An overriding concern for the safety of the hostages and for operational security made it imperative that we preserve as much secrecy as possible. We only went public today when it was clear a number of media outlets were preparing to report on the operation and that we would have no choice but to acknowledge it."[107]

Part of the secrecy stemmed from the fact that the mission failed. The hostages were never found. James Foley was beheaded on August 19, 2014. Steven Sotloff, the American–Israeli journalist also held by the group, was beheaded two weeks later.

Coalition spokesmen repeatedly told the press that they were doing everything possible to secure Moaz's release. But privately Jordanian officials sensed a lack of urgency on the part of the allies. King Abdullah wanted CTB-71, *his* unit, to be involved, as well. But intelligence that was both accurate and actionable was key in launching a successful strike and collecting real-time information inside Raqqa was near impossible. The city was primed and on full alert expecting some sort of retaliation or rescue attempt. Roads leading in and out of the city were booby-trapped with powerful IEDs; defensive positions were reinforced with squads of gunmen, particularly Chechens and North Africans, because of their combat experience in recent fighting, who were positioned with antiaircraft guns and even shoulder-fired missiles to repel any lightning-fast attempt to land commandos near the Islamic State's capital.[108]

King Abdullah ordered the GID to assemble a special task force to identify and locate the men who held Moaz. The GID believed that if they could find those responsible, they could also pinpoint where Moaz was being held. "We believe that he was moved every day while in custody," Colonel Mohammed,* a senior officer in the GID's counterterrorism division, explained.[109] "When Moaz was moved, it was primarily done under the cover of darkness," Colonel Mohammed elaborated. "He was moved with great concealment, and always under heavy guard."[110]

The GID relied heavily on human sources, primarily agents they recruited in the field. But the service's assets inside Raqqa were unable to get close enough to anyone inside the Caliphate security apparatus to pinpoint Moaz's location. Anyone suspected of helping the enemy was tortured and executed by the Islamic State's aggressive and highly capable internal security organs. There were summary executions throughout the territories controlled by the Caliphate; most of those killed were innocent, of course, informed on to settle a score, but the counterintelligence thugs always believed it better to err on the side of caution. "The Islamic State gunmen holding Moaz were incredibly paranoid, more paranoid than they were in the day-to-day running of their war. They were certain that there would be some sort of spectacular operation mounted to rescue the pilot and they moved him constantly," a GID officer who identified himself by the pseudonym of "Walid" explained. "Real-time intelligence-gathering from sources and assets in the vicinity was virtually impossible to ascertain. The moment we received a lead, we were always late."[111]

---

* A pseudonym.

Coalition intelligence services, primarily the CIA and NSA, relied heavily on SIGINT and ELINT measures. The United States also provided satellite imagery, reconnaissance overflights and even drone missions over the city. The effort revealed very little, however. Islamic State operatives displayed a keen knack for counterespionage tradecraft. Islamic State military commanders changed SIM cards on their mobile phones frequently; they chatted on encrypted applications, such as the Russian-designed Telegram.[112] There was very little chatter, and by the time a hint of communications could be detected, the intelligence was old and no longer useful. The spies—Jordanian and American—had a grudging respect for the tradecraft displayed by their adversaries in the Islamic State.

Quick action was imperative. The longer Moaz languished in the custody of the khawarij, the outlaws of Islam, as King Abdullah referred to them, the less likely it was that he could ever be retrieved.

There were discussions and even plans drawn up at JSOC and coalition headquarters to launch a rescue attempt. The GID and the JAF, as well as CENTCOM and intelligence arms of the In-herent Resolve coalition found it virtually impossible to gather real-time and reliable intelligence on where Moaz might be held. Launching high-risk top-tier special operations missions without the required intelligence was neither tactically nor strategically prudent. With the invaluable and irrefutable intelligence, the rescue operation never materialized.

7

Barbaric

Man's inhumanity to man / Makes countless thousands mourn!
—Robert Burns, *Man Was Made to Mourn: A Dirge*, 1785

THE ISLAMIC STATE might have advertised itself as a divine-inspired reincarnation of an Islamic empire from the Middle Ages, but it was run like any true twentieth-century despotic dictatorship. Abu Bakr al-Baghdadi was, of course, the supreme leader. He was very much like Saddam Hussein since his word was indisputable, his wisdom unprecedented and his power maintained courtesy of an inner circle of ruthless sycophants who were too preoccupied with undercutting one another to ultimately plot against him. The ruling council, the Majlis Shura al-Mujahidin, was made up of heads of the many Islamic entities fighting in Iraq and Syria who ran the day-to-day operations of the Caliphate—from garbage collection to planning spectacular catastrophic overseas attacks. They resembled the Soviet politburo more than a roundtable of holy men, and bureaucracy ruled the day. This most powerful legislative and decision-making body was responsible for the administration of the territories inside the confines of the Islamic State, its economy, its foreign policy, and of course its military operations and international presence, including operations around the world. In theory, the Council was all-powerful and, in fact, had the power to fire al-Baghdadi if they determined that he was acting against the Islamic principles and fundamentalist provisions of the Caliphate.[113]

The Majlis Shura al-Mujahidin was the brainchild of Abu Musab al-Zarqawi, the forefather of the Islamic State, who formed the original council in 2005 for his own terrorist army. The Islamic State followed the path of its founding father and expanded the governing body to the new entity that swept across Iraq into Syria; Abu Bakr al-Baghdadi was one of its founding members and ultimately its head. There were eleven members who sat on the roundtable—nine were selected in 2010 as the underground movement redefined its political and religious vision. The council members received rewards for their positions as decision makers. Each earned kickbacks from the extortion and oil revenues earned from the seized territories; as a result, the tribes, families and friends of those on the council earned power, prestige and wealth, and these perks warranted violent and unconditional loyalty.

Most of the leaders on the council were not clerics at all but rather former security and political men inside the Baath Party incubator of political ruthlessness that was Saddam Hussein's rule over Iraq. Abu Mohammed al-Adnani was al-Baghdadi's favorite. Like al-Baghdadi, he was ultimately apprehended by coalition forces and imprisoned at the notorious Camp Bucca; he was released in 2010 and hooked up with al-Baghdadi. He rose up the ranks of command with meteoric speed to become the second-most powerful man inside the Caliphate. He was undoubtedly the most feared. He copied Zarqawi's dress—black robes and combat gear—and his mentor's appetite for murder. Yet al-Adnani was different. He relished the carnage and understood the theatrical value of cold-blooded cruelty. CIA and GID officers labeled him as both the tactical and marketing genius behind the Islamic State's march across the region and its ability to sow fear in its enemies and attract volunteers from the four corners of the world.[114]

Al-Adnani oversaw three of the Islamic State's most important organizations. He commanded a global network of compartmentalized sleeper terror cells throughout Europe and elsewhere in the Middle East. These men and women underwent extraordinary military training, complete with extensive background checks and loyalty tests before they were sent back to their home countries.[115] Al-Adnani commanded the Emni, the Islamic State's much-feared internal security counterintelligence secret police. The Emni gave al-Adnani absolute power in eliminating suspects and labeling political rivals as spies. The Emni was a judicial end unto itself—trying, torturing and executing whomever it deemed an enemy to the Caliphate. And al-Adnani was the Islamic State's marketing mastermind behind the organization's fearsome image. He approved all the Twitter messages and he choreographed the ghoulish beheading and execution videos posted to YouTube. He personally approved the scripts, the musical score and the backdrops for the executions.

The ruling council decided to kill Moaz shortly after he was pulled out of the Euphrates. His death sentence was voted on and passed unanimously. The members of the council argued that the Jordanian pilot was worse than a Crusader—he was a Sunni bombing the true believers in the new Caliphate—and, as a result, his death had to be epic in its shock value; mercy, they feared, would be a sign of weakness. It was imperative that the Jordanian pilot's death destroy the Arab resolve for the war against the Islamic State. Al-Adnani knew that there was Madison Avenue messaging in not only the act of cold-blooded murder but in its cringeworthy barbarity. Al-Adnani had an idea of a spectacle that was so horrifically game-changing that it would force the Jordanian people to turn against their king. He told the council his plans. It was medieval, but no one dared object. Al-Adnani was, after all, speaking for the caliph.

Shortly after his capture—and in between the beatings— Moaz was brought into a dark room somewhere underneath a building in Raqqa and seated behind a large wooden table facing a video camera on a tripod. Spotlights were positioned to face him, as well as to provide soft illumination. A cameraman, several technicians and al-Adnani supervised the filming in the underground dungeon-like room; a soundman checked audio levels. There were no attempts to conceal the bruises around Moaz's face; his left eye looked as if it had been subjected to several rounds in the ring with a heavyweight champion boxer. On cue, Moaz was ordered to reveal details about the coalition and the Arab participants. He listed which Arab countries were involved in the fight against the Islamic State and what aircraft they were flying. He was next ordered to confess, and in full detail he listed the A-to-Zs of his mission on the bombing run of Raqqa. The confession should have saved his life, but the Islamic State was not known for benevolence. Moaz was returned to his cell. The beatings continued, as did the interrogations.

On the morning of January 3, it is believed, Moaz al-Kasasbeh was dragged from his holding cell and brought to an area of Raqqa that had been hit hard by coalition air strikes. The area, on the northeastern fringe of the city near the central prison and the main football stadium, sat on the strategic Road 6 and, as a result, was an Islamic State transportation hub.[116] And, as such, the area was a primary target for coalition air strikes. Kasasbeh's face was unshaven and puffy. His lips and eyes were swollen from the beatings he was subjected to; he walked with a pronounced limp and wore a pair of secondhand leather shoes. Moaz was issued with a new orange prison blouse and trousers. A dozen heavily armed Emni enforcers surrounded him.

Technicians laid electronic and sound cables throughout the area. They prepared the area as an impromptu on-location film set amid the bombed-out ruins. It was a brilliantly sunny day. Abu Mohammed al-Adnani was in his element, playing master propagandist and fledgling film director. He insisted that the footage was shot in the early-morning sun so that the natural light of the new winter's day could be enhanced by the illumination from the portable stadium lights the media department purchased. A guard of Islamic State special operations personnel was already on the set. Their desert camouflage uniforms were right-out-of-the-box new. Sand-colored balaclavas covered their faces, and their AK-47 assault rifles were strapped over their chests aimed downward. A few dozen men were positioned to stand side by side in a long line that led from the rubble through a clearing toward an iron cage that was three meters high and three meters wide; others had climbed the blown-out shell of an apartment building and stood at attention inside the blast-created cutouts.

Al-Adnani selected the set, positioned the extras and made sure that his star was prepared in wardrobe. Moaz was ordered to walk past the bombed-out buildings slowly; he was told to look up and scan the wreckage, raising his head toward the devastation and then to the heavens. The director told him to appear contrite. Cameras followed his every move and from various angles. The media crew consisted of five men with four cameras.[117] Some of the cameras were on heavy tripods, like those used to film Hollywood productions.

Moaz moved sluggishly, appearing as if he was slightly sedated. He was ordered into the metal cage and ordered to stand and face the men with guns. As he stared, two Islamic State gunmen poured gasoline on him and on the floor of the cage. The young pilot appeared paralyzed by the fate about to befall him. His bruised faced showed helpless confusion. He bowed his head in prayers, asking God to make his suffering end mercifully.

The cameras rolled. The privilege of executing a prisoner was usually handed to one of the foreign fighters, or one of the burly fanatic Chechens that reveled in bloodcurdling sadism delivered to enemies of the Caliphate. But al-Adnani wasn't about to deny himself the climax of his hard work. The GID is confident that al-Adnani changed out of his characteristic black robes and donned camouflage fatigues to look like all the other gunmen. Al-Adnani grabbed a long wooden torch and dipped it in the trail of petrol. A fast-moving wall of fire raced to the cage and engulfed Moaz in seconds in a ferocious wave of flames. Moaz writhed in agony, attempting in vain to fight off the fast moving incineration. The orange fire and the black acrid smoke was overwhelming. Moaz was dead within a minute. The lifeless charred body dropped to its knees and then fell backward. The suffering he endured during the murder was simply unimaginable.

Al-Adnani ordered one of the cameramen to go near the cage and grab a closeup of Moaz's charred corpse. The camera focused on each piece of the body that had been burned in the ferocious immolation. Satisfied that he had the footage he needed, al-Adnani summoned a red bulldozer that was on standby. The earthmoving machine dropped rocks and sand onto the fiery remnants of the cage and rolled back and forth over the debris, flattening it in one final desecration by men of boundless cruelty. The barren lanes were cleared of Islamic State fighters and any other trace of the horror performed on it. Al-Adnani was concerned that coalition drones flying overhead might learn of what had just transpired and that air strikes or even a commando raid would interrupt his project.

The Islamic State had just staged and filmed its most potent act of horror to date, a permanent message of terror, challenging the world to try to match the Islamic State in battle. But there was still a lot of work to do: a soundtrack had to be added, and there were special effects and graphics to insert in postproduction. The work was done in great secrecy. The raw footage and the files were kept on computers that were not connected to the web to safeguard the system from coalition hackers. Emni guards and a legion of Islamic State commandos protected the basement media labs where the computer whizzes—young men whose talents could have easily earned them a healthy income in Hollywood—plied their special skills on state-of-the-art computers. The Islamic State was determined to release the shocking video to the world at a time of its choosing.

8

The Extortionists

He is a barbarian, and thinks that the customs of his tribe and island are the laws of nature.

—**George Bernard Shaw** (*Caesar and Cleopatra*)

THE KING RECEIVED a daily intelligence briefing every morning from the GID. Urgent information from the spies was delivered at all hours. On the morning of January 8, the GID received word from a source inside Raqqa that Moaz had been executed, though the details of the method used to kill him were still unknown. The source was one of the GID's most reliable assets in the Caliphate.

Days after Moaz's reported murder, a senior officer in the GID Counterterrorism Division, the CTD, received a call on his mobile phone. The caller ID was identified as "Unknown" and the voice on the other end spoke with a muffled voice. In a North-African-accented Arabic he identified himself as a ranking member of the Islamic State. The GID officer wanted to engage the caller and hoped that the call could be traced and the man's location triangulated to coordinates inside Syria, but the North African kept it short by making a singular demand. "We want sister Sajida," the caller barked before hanging up.[118]

The insertion of "sister Sajida" into the crisis was both troubling and puzzling. Sajida al-Rishawi was a forty-six-year-old failed Iraqi suicide bomber from Anbar Province on death row in Jordan. She was illiterate and paunchy. Her first husband, a trusted lieutenant of Abu Musab al-Zarqawi's al-Qaeda in Iraq, was killed in the hand-to-hand battle for Fallujah in 2003. Two of her brothers were also killed fighting American forces. Al-Rishawi was supposed to have blown herself up along with Ali Hussein Ali al-Shamari, her second husband, in the Radisson SAS hotel in Amman on November 9, 2005, as part of a three-pronged suicide bombing attack in the Jordanian capital. But al-Rishawi's suicide vest was defective, and when she failed to detonate the device, her husband scolded her and ordered her out of the hotel lobby. He then walked into a wedding reception being held for close to a thousand guests and pulled the toggle switch on the explosive device he wore under his coat. Thirty-six people were killed in the bombing and dozens more were wounded. Suicide bombers hit two other hotels that night—the Grand Hyatt and the Days Inn.

In all, sixty people were killed in the Amman hotel bombings and over one hundred were injured. Because of the steep death toll, the attacks are considered Jordan's 9/11.

The GID arrested al-Rishawi shortly after the bombings. She confessed her part in the attacks on Jordanian national television and displayed how she wore her explosive device under her clothes. A Jordanian court convicted her of terrorist crimes and sentenced her to death. She languished in an isolated cell in Juweidah Women's Prison near Amman for nearly nine years, seen only by her court-appointed lawyers. Al-Rishawi was virtually forgotten inside Jordan and beyond. She was never mentioned in any of the al-Qaeda in Iraq and then the Islamic State propaganda.[119]

The GID did its part to make sure that references to Sajida were kept out of the press. But now Sajida gave the Islamic State a new talking point, and it became a means to weaken Jordanian resolve in the war. Her inclusion in the process was a subtle reminder that the Islamic State had reached two of the country's international borders and a threat that Jordan was next in the Caliphate's sweeping march across the Middle East.

The foundation of the Islamic State's economy was anchored in criminal profit—along with the vast reservoirs of stolen oil and cash, hostages were a valuable currency. The international fuss around the men in their custody who were forced to wear orange jumpsuits while a blade was held to their jugular was magnified by the capture of Lieutenant Kasasbeh. The Islamic State's Byzantine human bazaar was open for business.

Two Japanese nationals—contractor Haruna Yukawa and journalist Kenji Goto—were prisoners of the Islamic State. Negotiations to buy the pair's release were unsuccessful. Emissaries from Raqqa were demanding a king's ransom for the men, and the Japanese government, key financiers of the Inherent Resolve coalition, in addition to providing $150 million in humanitarian aid to Syrian refugees in Jordan, refused to meet the terrorists' demands.

In early January 2015, Twitter accounts associated with Islamic State followers posted a video of the two Japanese hostages in orange jumpsuits, on their knees. They stared into the camera with blank hopelessness. Jihadi John gyrated behind them as if he were a DJ at a rave. He was dressed in black with tan desert boots and a leather holster strapped across his left arm; the barren expanses of the northwest Syrian desert served as backdrop and soundstage. The dagger he had used to behead other Islamic State prisoners was held firmly in his left hand. The video was called "A Message to the Government and People of Japan," but his diatribe, delivered with a forceful London street accent, was nothing more than pure mobster extortion. "To the prime minister of Japan: Although you are more than 8,500 kilometers away from the Islamic State, you willingly have volunteered to take part in this crusade," John said. He then demanded $100 million for each man's life, the same $200 million that Japan's Prime Minister Shinzo Abe had pledged to aid the people of Syria in Cairo on January 17.[120] Jihadi John gave the Japanese premier exactly seventy-two hours to acquiesce and deliver the cash, otherwise, John warned as he gestured with the dagger in his hand, "This knife will become your nightmare."[121]

A new video surfaced days later. This time only Goto appeared. He held a photo of himself sitting next to the decapitated corpse of Haruna Yukawa. A narrator with a poor Japanese accent blamed Prime Minister Abe for the hostage's murder and added that unless Sajida al-Rishawi was freed, Goto would be killed.

The GID could no longer conceal the Sajida factor. Her name, and the painful reminder that Zarqawi was the architect of what was now the Islamic State, had taken center stage in the global arena.

Governments, as a rule, do not negotiate with terrorists. The United States *did not* negotiate with terrorists. At least officially it didn't. That was the policy of administrations since Ronald Reagan took the oath of office in the wake of the Iran hostage crisis, but the reality was once described as more Hollywood than history.[122] The stoic resolve was great for press conferences and the campaign trail, but the United States did, indeed, negotiate and barter with terrorists, and it didn't matter if the president was a Democrat or a Republican. President Ronald Reagan traded arms for hostages with Iran and Hezbollah and years later secured the release of American captives by outlaws and declared terror factions all over the world—from South America and Southwest Asia.

Officially the Hashemite Kingdom of Jordan did not negoti-

ate with terrorists. But the GID *did* speak to terrorists—all espionage services had contacts inside terrorist organizations, and
they all maintained dialogues with entities that their nations
were at war with. The spies followed a Michael Corleone view
of the world by keeping their friends close and their enemies
closer, but they didn't barter with them. "It isn't the policy of
this organization to strike bargains and compromises with mur-
derers and kidnappers," Colonel Ahmed,* a twenty-year veteran
of the CTD, stated. "But," he added, "there were always ways
to communicate."[123]

The GID demanded that the Islamic State provide proof that
Moaz was still alive, knowing, of course, that the pilot was already dead. None was offered because none could be. The men
in Raqqa raised the ante once again.

On Tuesday, January 27, the Islamic State released another
video on Caliphate-connected Twitter accounts. The brief clip
showed Goto, though this time he was holding up a photograph of
a bearded Moaz al-Kasasbeh. "I only have twenty-four hours left
to live, and the pilot has even less," Goto said. "Any more delays
by the Jordanian government will mean they're responsible for
the death of their pilot, which will then be followed by mine."[124]

The next day a new Jihadi John video clip appeared on Islamic State–friendly Twitter accounts. The setting was pastoral,
with mountains and the green shrubs of winter in the desert. The footage, though, was gruesome. Jihadi John beheaded
Kenji Goto with savage glee. Everyone feared that Moaz would
be next. Candlelight vigils were held in Aiy, in Amman and
throughout Jordan. Thousands of eleven-by-fourteen prints
were made showing Moaz in an air force sweater and a gray-
blue beret. The mood inside the country was tense. Moaz al-
Kasasbeh's plight forced many to question Jordan's participation
in the international alliance against fellow Sunnis in a neighboring Arab land.

---

* A pseudonym to protect his identity.

Even members of the Kasasbeh clan began to speak out, questioning Jordanian participation in the effort to defeat the Islamic State. "When he joined the air force, I expected him to defend Jordan, but not to go fight in another country," his mother said. "This is not our business being there."[125]

There were elements inside the country that sympathized with the fundamentalist ideology, if not how they enforced Islamic laws and edicts. The mood in the country was worrisome to the United States government. The email traffic was fast and furious between the US Embassy in Amman and Foggy Bottom in Washington, DC. The situation, as a US security official based in the embassy would later describe, was "tense and delicate."[126] The White House wanted to keep Jordan stable and the coalition vibrant.

The saga appeared to be coming to an end with the painful outcome all but a foregone conclusion. But King Abdullah warned the Islamic State that their manipulation of lives would come with a price, and he wanted the GID officer who was in touch with an Islamic State representative who they believed to be North African to convey a message to the others in his group. "Tell him to ask the other Arabs in the terror army about the Jordanians," the king instructed the officer. "Make him understand that when you get on our bad side, we are coming after you. Make him tell his people that they can make it easier on themselves or they are going to see us, even if it means that the army crosses the border and marches all the way into Raqqa."[127]

The ruling council in Raqqa didn't know that the GID had already established that Moaz was dead. They wondered if they had overplayed their hand, especially when the message from Jordan was crystal clear: if Moaz was harmed, Sajida and others from Zarqawi's terror network who were on death row in Jordan would hang.

9

Shock and Horror

Terrorism is the tactic of demanding the impossible and demanding it at gunpoint.

—**Christopher Hitchens**

UP UNTIL JANUARY 2015, the bearded men speaking a Rosetta-stone catalog of languages fought fiercely to expand the Caliphate. Their advance was preceded by their barbaric reputations, and the territories they seized were savaged by medieval crimes and punishment; the message of the Caliphate was that it was an unstoppable military force that vanquished all in its way. And then came the battle for Kobanî.

Kobanî was a strategic crossroads that touched up against the Turkish frontier in northern Syria. The Islamic State captured most of the area around the city in September 2014, which sparked a massive humanitarian crisis when close to four hundred thousand people, primarily Syrian Kurds and Yazidis, fled in panic. But the city's Kurdish garrison of the city, primarily forces from the People's Protection Units, or YPG, didn't flee. They stood fast and made the militants pay a dear price for every street and alley they tried to secure. Thousands of Islamic State fighters laid siege to the city.

Abu Mohammed al-Adnani's public relations apparatus wanted to use the battle for Kobanî as a recruitment vehicle to attract foreign volunteers. Islamic State commandos, the Inghemasiyoun, Arabic for "those who immerse themselves,"[128] were filmed with their characteristic bright blue bandanas fighting a close-quarter, winner-take-all battle against the Kurds. The bearded men shouted "victory or martyrdom" as they waved black banners and loaded explosive vests onto their torsos. The videos and articles promised the world that victory was near.

But Kobanî wasn't Ramadi or Mosul. The city's defenders didn't abandon their posts, leaving behind the women, the children and the elderly. The defenders were reinforced. Kurdish militias from Syria and Turkey coordinated their actions to bolster the city's encircled garrison. Iraqi Kurds, led by the famed Peshmerga and units often led by women, joined in the battle, as well. The game changer, however, was the involvement of Inherent Resolve special operations units deployed behind Islamic State lines to support the Kurds.

Small teams of operators from the US Army 1st SFOD-Delta[129] and the CIA Special Operations Group were inserted into the area to train Kurdish forces and to conduct reconnaissance, surveillance, eyes-on-target intelligence and targeting missions. The allied special forces teams worked covertly, away from the reach of media cameras covering the conflict, and they played a pivotal role in enhancing the Kurdish zeal with special operations capabilities. Teams from 22 Special Air Service Regiment were at the forefront.[130] The SAS fought side by side with Kurdish forces, engaging Islamic State shock troops in pitched battles and at close range. The assaulting squads of jihadists and suicide bombers were no match for the SAS troopers. In some of the close-quarter melees, SAS units killed hundreds of terrorist targets. SAS snipers, legendary throughout the world's special operations fraternity, picked off target after target.

The British effort, and indeed much of the special operations campaign, was led by Major General Mark Carleton-Smith, the Directorate of Special Forces commander. Originally commissioned with the Irish Guards, Carleton-Smith spent many years in 22 SAS, including tours in Northern Ireland, Bosnia and Afghanistan.[131] Under his leadership the SAS was able to return to its roots of desert warfare from when the legendary Captain David Stirling's Long Range Desert Group fought in North Africa during the Second World War. The British operators called in air strikes that obliterated Islamic State efforts to reinforce their frontline forces in the city attempting to break through Kurdish lines. The men in long beards were forced to resort to terror tactics, such as sending suicide bombers into the fray, to try to regain the battlefield initiative, but SAS snipers and dedicated mortar fire eliminated the fighters with explosives strapped to their chests before they could get close to their targets.

Islamic State forces pounded Kobanî for four excruciatingly bloody months of relentless pressure. But the Kurds and other forces, including the Free Syrian Army and Syrian Democratic Forces,* held on with tenacious resolve. Airpower, though, was the difference. It took over seven hundred coalition air strikes, nearly three-quarters of the missions flown in the Inherent Resolve campaign, to dislodge the Islamic State garrison from the town.[132]

On January 26, 2015, pounded from the air and on the ground into a humiliating stalemate, Islamic State forces conducted a tactical withdrawal from the city rather than suffer a stunning battlefield defeat. Kobanî was the first major town inside the

---

* The Syrian Democratic Forces (SDF) is a multiethnic and politically diverse alliance of pro-Western anti–Islamic State and anti-Assad Kurdish, Arab and other minority group (Assyrian, Chechen, Turkish and Yazidi) militias that is usually led, at the front, by Kurdish officers.

borders of the Caliphate to be liberated. Kurdish and Free Syrian Army soldiers danced jubilantly before news cameras, stomping on the black flag of the Islamic State.

The Islamic State's battlefield defeat was a public relations nightmare for Abu Mohammad al-Adnani. He realized that news networks all around the world were broadcasting Kurdish fighters celebrating their win with the *chapi*, a circular hand-holding dance, enjoyed in festive occasions. Kurdish fighters were only too happy to tear Islamic State flags in half and to spit on them. The Majlis Shura al-Mujahidin council met to discuss a response. The men understood that the combination of allied air power and special operations units on the ground was a battlefield game changer. Their forces did not have the capabilities to launch new offensives like what they mounted in Iraq earlier the year before. Al-Adnani reported that his sleeper cells in Western Europe, primarily in Brussels, were not yet fully trained or financed. The Islamic State special operations chief realized the importance of recapturing the initiative and redefining the narrative. The filmed immolation of Moaz al-Kasasbeh had now become a strategic tool. The gala premiere of the Moaz production was scheduled sometime in February.

It was an early wake-up call at the Secret Service and State Department's Diplomatic Security Service field offices in Washington, DC, on the morning of February 3, 2015. The federal agents arrived at Joint Base Andrews well into the middle of the ungodly hours of nighttime and assembled at the VIP building before dawn. Two armored limousines and a small armada of black Suburban follow cars were already waiting, as were several members of the Jordanian Royal Guard, one of the special operations branches of the Jordanian military entrusted with protecting the royal family and its institutions. When members of the Royal Guard were in Jordan, they wore dazzling digital-pattern camouflage fatigues and distinctive plush green berets, but on the tarmac one cold morning in Prince George's County, Maryland, the men wore finely tailored business attire, their semiautomatic pistols tucked neatly into holsters concealed by their dark blazers, as they awaited the arrival of their king.

The king's trip had not been on the schedule, and the advance elements of his protective detail were ordered to Washington with less than twenty-four hours' notice. The Jordanian agents rubbed their eyes and yawned, battling jet lag and the February shivers. The team had already visited the king's hotel to review the security precautions, and they conducted site inspections of the locations on the monarch's itinerary. The men looked at the mobile phones and checked their radios. When they heard that the aircraft was a few minutes out, they placed their earpieces in and grabbed one of the doughnuts that the US Air Force made sure were on hand for every VIP arrival. The plane landed shortly after sun's first light.

This was King Abdullah's second visit to the United States in three months. King Abdullah came to Washington to discuss Jordan's urgent need for American military and humanitarian aid. The situation on the battlefront and regarding Syrian refugees was dire. The king was accompanied by his foreign minister, Nasser Judeh, who was to sign a memorandum of understanding, or MOU, with US Secretary of State Kerry later that morning to enhance the bilateral strategic relationship with increased American aid. The fate of Lieutenant Moaz al-Kasasbeh hung heavy over the planned meetings.

The security-package motorcade traveled at top speed to the Four Seasons Hotel in Georgetown. Jordanian and American flags were already flapping in the wind awaiting the arrival of the state visit. Staffers were already working in the Presidential Suite and the adjoining rooms preparing communications links to Amman and fine-tuning a hectic itinerary. King Abdullah settled in and reviewed some notes he wanted to share with the senators and congressmen on Capitol Hill later in the day.

Half a world away in Syria, it was already close to dusk. The black smoke plumes rising from Islamic State military positions hit by the clockwork launch of coalition air strikes disappeared into the oncoming darkness. In Raqqa, hit hard that day by the bombing runs, the media center was busy at work preparing to transmit its snuff film to the world. News of King Abdullah's visit to the United States gave Abu Mohammed al-Adnani his perfect media moment for the international debut of his cinematic masterpiece and to eclipse the defeat at Kobanî. He knew that it would be a game changer.

Secretary Kerry arrived at the Four Seasons Hotel shortly after breakfast and met with King Abdullah to discuss developments in the war against the Islamic State. Ambassador Wells accompanied Kerry. There were smiles for the camera and a photo op that included Crown Prince Hussein, the king's twenty-one-year-old son and heir to the throne. Jordan carried out more air strikes than any of the other non-American coalition partners, and it could not continue the sorties without more American assistance. Period. The MOU signing was a nice gesture, but it was as binding as a promise that the check was in the mail.

Secretary Kerry and Foreign Minister Judeh signed the MOU in a small and hastily put-together ceremony held in the hotel's banquet room. Both men smiled as the United States pledged aid to Jordan from $600 million to $1 billion annually—an increase to help offset the Herculean costs of taking care of the 1.3 million Syrian refugees inside Jordan and to help finance Jordan's war on terror. At the brief press conference that followed, Secretary Kerry reminded the world of the Jordanian prisoner. "The people of Jordan need to know that all Americans will join them in praying for the early and safe return of Lieutenant Moaz al-Kasasbeh," Kerry stated in a sharp voice. "And we call upon his captors to release this brave man so that he could return to his family and his homeland, to at least provide proof of life, which Jordan has asked for."[133]

King Abdullah didn't hear Kerry's remarks. His motorcade was already racing down Constitution Avenue toward the congressional offices. The king was silent on the ride to the Capitol. A staffer led the king to the Arizona senator, who greeted the king at the entrance to his office. As head of the Armed Services Committee, McCain was one of the most powerful men in Washington. They waited until the doors were closed before talking. It was 11:25 a.m.

A world away, Mohammed Abu al-Adnani sent an SMS to an aide authorizing him to instruct the media center to tweet his tour de force. The film was titled "Healing the Believers' Chests" and was precisely twenty-two minutes and thirty-four seconds long. The dark of night already covered Raqqa and the sounds of explosions were heard in the distance, but al-Adnani's media machine was busy. Screens were quickly set up throughout Raqqa to show the gruesome video on a continuous loop. The macabre spectacle attracted an audience of thousands. Islamic State gunmen watched Moaz's barbaric execution as an Islamic State commander, a Libyan, whipped the crowd into a frenzy chanting, "God is Great!" and "Death to Jordan!" There were always cries and chants demanding the death of America. Parents brought their children to the town square to watch the footage. Vendors sold grilled meats and soft drinks in what soon became a carnivalesque atmosphere. Media center cameramen made a propaganda video of young children cheering as they watched Moaz burn.[134]

It was evening in Amman. Families around the city were gathered around the dinner table, and those heading home from work negotiated the twisting and often maddening traffic that clogged the city's main arteries. The lights were still on inside the GID headquarters campus, where case officers on the Moaz Task Force were about to call it a day. And then the tweet hit. Phones in the office began to ring incessantly. Mobile-phone messages were coming in at a furious pace. Veterans of the country's shadow wars sat stone-faced and in shock.

An analyst assigned to the CIA's Amman Station told one of his counterparts in the GID that the office went from "Silent Night" to "Times Square" in an instant.[135] The emails were coming in faster than the evening crew could respond. Everyone in Langley was still at their desk—it was too early to go and grab lunch at the headquarters commissary.

In Kuwait, at Camp Arifjan, CENTCOM and coalition officers watched the tweet repeatedly, finding the footage hard to watch and harder to believe. Years of combat experience and staff college study had not prepared them for such unabashed cruelty. Perhaps Marine Corps General (Ret.) John R. Allen, President Obama's special envoy for the Global Coalition to Defeat ISIS, summed it up best when he said, "We were seeing atrocities, more horrific than any I have ever seen or even could have imagined: the beheadings, the crucifixions, the electrocutions, the drownings, and of course the one that I believe focused the collective horror and rage of the world."[136]

The outrage was but one of the objectives of the video. Al-Adnani and his propaganda specialists in Raqqa wanted to make every coalition pilot think twice before embarking on combat runs over the Caliphate.

The pilots in No. 1 Fighter Squadron were preparing the next day's F-16 sortie at dawn when news of the tweet hit headquarters. It had been forty-one days since Moaz was captured, and the mood in the squadron was grim. The older pilots, men of the senior staff, had yet to commit their personal lives to social media and weren't connected to Facebook or Instagram and didn't know what Twitter was. The young Viper drivers were the first to hear of the video, and they took the news the hardest. The secure landlines began to ring minutes later, as did personal mobile phones, each with a distinctive ringtone. The pilots muted everything. They found a corner to watch the tweet alone and in private. "We didn't know what we were about to see," one of the junior pilots confessed. "We knew it was going to be awful."[137]

The narration of "Healing the Believers' Chests" was entirely in Arabic. The production was specifically directed at the Sunni countries of the region participating in the anti–Islamic State coalition. The film's style was slick, though choppy, with frequent cutouts and image juxtaposition. One of the RJAF officers commented that it was designed to resemble a military episode on the History Channel. The opening segment was nothing more than a montage of spliced stock footage showing US and Jordanian cooperation in the war on terror—it was Propaganda 101, a diatribe of anti–King Abdullah statements meant to spark sympathy inside Jordan. The messaging was cut with images, stills and video of burned victims of coalition air strikes against Raqqa and other Islamic State targets; babies and children burned to death were highlighted. The gut-punching horror came next.

Using Google Maps, "Healing the Believers' Chests" showed the location of all the coalition air bases used to bomb the Islamic State. Then Moaz appeared, dressed in his orange jumpsuit from when he was filmed in the basement in Raqqa. His face was bruised. Speaking in a calm and deliberate Arabic, he revealed the secrets of his squadron and the RJAF's work with coalition forces. Using video images extracted from Moaz's mobile phone, the film featured clips of actual pilot briefings inside No. 1 Squadron, revealing the identities of the personnel. Moaz then detailed which Arab countries were flying which airplanes; which six Arab nations—Jordan, Saudi Arabia, the United Arab

Emirates, Bahrain, Qatar and Morocco—were executing air strikes; and which Arab states—Kuwait and Oman—were providing support and bases of operations. The film then continued to Moaz's execution: the walk through the rubble in front of the Islamic State gunmen, the cage and then the immolation. A soundtrack of Islamic chanting was inserted to accompany the gory footage of the young pilot burning.

Some of Moaz's comrades wondered if he had been drugged. Some hoped that for his sake he was. "I think that he knew what was going to happen to him and he was deep in prayer," said Lieutenant Hani,* one of Moaz's closest friends in the squadron. "He was a very religious young man and he would have focused his thoughts on prayer."[138]

The pilots of No. 1 Squadron wept openly. Men in flight suits sat on the wooden benches inside the briefing room with their heads held low and their hands wiping an unstoppable flow of tears. They simply couldn't fathom the barbarity used to kill their friend; other pilots simply threw their phones down in disgust, unwilling and incapable of watching any further. But as disgusted as they were watching their friend and comrade burn to death, the ending of al-Adnani's vitriolic film warranted the greatest concern. The Raqqa Media Center had taken the photos of the squadron's pilots on Moaz's phone, along with their personal information, and included them in the film's closing credits. The words "Wanted Dead" were placed above each of the pilots showcased. The film offered one hundred gold dinars to anyone who killed a Crusader pilot.

The pilots wanted revenge—they wanted a mission. "We want to bomb the hell out of them," one of the younger pilots demanded. "We can't let them get away with this." Lieutenant Colonel Ali did what he could to check the emotions of his fliers, but he knew that the gut instinct was vengeance.[139] Ali ordered

---

* A pseudonym.

his men back to work. There was a sortie the next morning. Then he shut the door to his office and wept in silence.

Senator McCain and King Abdullah were behind closed doors when an aide to the Jordanian monarch knocked on the door and entered. The king knew that something was amiss; staffers knew better than to interrupt such important meetings. The aide whispered in the king's ear in Arabic, saddened that he had to be the messenger of such bitter news. Senator McCain also knew something terrible was afoot.[140]

The king's reaction to the tweet was outrage, but the Senate was not the place to punch a wall in anger. The king realized that he had to temper his rage. "The King did all he could to contain his anger and calculate the ramifications of the footage," Akel Biltaji, a former mayor of Amman and trusted advisor who was traveling with the monarch, recalled. "He knew that he had to return to Jordan at once, but he also knew that it was imperative to keep his scheduled meetings on Capitol Hill."[141] The monarch summoned one of his aides and gave him a list of tasks, including to get the aircraft ready and to have everyone at Joint Base Andrews by evening. It was imperative that he return to Jordan immediately.

One of John McCain's staffers had, in the interim, made sure that the senator was apprised of the Islamic State tweet. "McCain was no stranger to torture and the mistreatment of prisoners of war," the king recalled, "but there was nothing really to say. What could anyone say to such an affront to humanity?"[142] McCain asked what he could do, to which Abdullah replied, "I'm still getting only gravity bombs, and we're not even getting resupplied with those. Meanwhile, we're flying 200 percent more missions than all the other coalition members combined, apart from the United States."[143]

Jordan needed smart bombs, laser-guided munitions and

stand-off air-to-ground missiles. Ammunition stores were dangerously low on vital supplies. The king argued that it was impossible to fight a twenty-first-century aerial campaign with ordnance that was rejected as obsolete during the first Gulf War twenty-three years earlier. Jordan's internal security forces needed US assistance to be able to meet the savage intentions of the Islamic State, as would its conventional military component.

Jordan's depleted arsenal was the primary reason that the king decided to keep his lunch meeting with members of the Senate Foreign Relations Committee. The king's hosts, senators Bob Corker from Tennessee and Bob Menendez from New Jersey, smiled for the official Senate photographer, but King Abdullah wanted to get down to business the moment the doors closed, and he wanted the senators to be on his side. It was hard to refuse the king's pitch, especially mere minutes following the tweet of the Islamic State video. The senators pledged their support, as did those on the Senate Appropriations Committee. Committee chairman Thad Cochran, a Republican from Mississippi, and Patrick Leahy, the long-serving Democrat from Vermont, promised the king that the Senate would do whatever it could to help Jordan. The king hoped that it wasn't just more lip service. His country's position as a frontline combatant in the war against the Islamic State was in peril, and that threatened the international coalition efforts and American wishes.

A blustery February wind accompanied King Abdullah as he walked out of the Capitol to his awaiting motorcade. The winter sun had already set over the Beltway. His aircraft at Joint Base Andrews was fueled, and members of the entourage were already in the VIP hall, enjoying a coffee, awaiting the king's arrival, but the Jordanian monarch had one final stop to make. President Obama was already waiting.

President Obama and King Abdullah disagreed over many

aspects of American policy in the Middle East: the 2011 withdrawal from Iraq and the overtures toward the Iranians. The
American presence and the shield that came with it prevented
other nefarious players from inching the Middle East closer to
all-out conflict. A withdrawal of the United States from the
area—especially from fires they helped light—threatened the
national security of many of America's allies.

President Obama knew that Inherent Resolve's alliance was
anchored in an unstable foundation. The "Moaz factor," as it
was known in some US military circles, proved just how tenuous the international task force was. Some air forces officially
halted their participation in combat air operations over Iraq and
Syria;[144] flights were called off after so little effort was made
to rescue the Jordanian pilot. But Moaz's capture personified
the lack of close combat search and rescue assets that American
forces were willing to dedicate to the international coalition.
There were real worries concerning negligent shortcomings in
how the coalition was protecting the pilots flying the round-
the-clock combat missions against the Islamic State, specifically
why CENTCOM had not positioned suitable special operations
and aviation assets in northern Iraq to rescue downed pilots.
Some coalition partners indicated that until special operations
forces with Osprey V-22 aircraft were based in northern Iraq,
ready to respond in seconds to another Moaz, they would halt
combat operations.[145]

Some coalition officers, including members of the US military, told their Jordanian counterparts in confidence that had
the captured pilot been an American or a Brit, the human drama
could have shattered the multinational effort.[146] It was fourteen
years since 9/11—the West was exhausted and any excuse to
pull out of another Middle East war on terror would have had
critical political implications. President Obama knew this. His
apprehensive approach to inserting American forces into a con-

flict that had no exit strategy created doubt at a critical cross-roads where determined leadership and all-out commitment was warranted.

A marine sentry in dress blues snapped a sharp salute to the Jordanian monarch as he entered the West Wing shortly after 6:00 p.m. Members of the White House press corps were invited to take pictures and to try to ask questions of the two leaders, but both men were silent and appeared uncomfortable. They sat and managed to forcefully produce smiles as the shutters snapped and the cameras rolled. Reporters shouted their questions but there were no replies. The one-on-one discussion commenced once the press was escorted out of the Oval Office. The king sat to the president's right. Both men had their hands clasped, knowing that they would be discussing bad news and the response that the most heinous murder warranted once the reporters left the room.

President Obama looked at the king and saw the concern and the exhausted exasperation in his eyes. He expressed his disgust and outrage, and, on behalf of the American people, offered his sincere condolences. He said that America was doing all it could to help Jordan. King Abdullah gave a list of weapons and supplies that the armed forces needed. He was reported to tell Obama that Jordan had three days of bombs left in its arsenal. Jordan was going to all-out war and he was going to use every one of them until they [the Islamic State terrorists] were gone.[147]

The king told the American president that Jordan had to act decisively—either unilaterally or not, and not just on the conventional battlefield. The effort of the war against the Islamic State had to refocus itself on eliminating the men who ran the show. It wasn't enough to launch air strikes against arms dumps and parking lots. The leadership, the irreplaceable leadership who emerged from the Iraqi security services and the Syrian prisons, had to be eliminated. This was the only way that the war would end.

There was one more thing, the king added. The men who were responsible for this affront to humanity—the ones who pulled Lieutenant Kasasbeh out of the Euphrates, beat him, and whose own inhumanity was veiled by pious misinterpretation—had to die. He would do it with American assistance or without. It was a national imperative.[148]

President Obama's foreign policy was based on risk aversion, but it was impossible to base moves on strategic calculations in a region where two plus two didn't equal four. There was genuine concern in the White House that unilateral Jordanian action could be a wild card in the war against the Islamic State and, should it backfire, require the involvement of large numbers of boots on the ground. Obama had no choice but to support King Abdullah's position and to pledge full American cooperation in any covert moves to go after Moaz's killers. The details would be left to the generals and spymasters to work out.

King Abdullah did not speak to reporters as he departed the White House. The only statement that emanated from the one-on-one meeting was from the National Security Council spokesman: "The President and King Abdullah reaffirmed that the vile murder of this brave Jordanian will only serve to steel the international community's resolve to destroy ISIL."[149]

The king didn't have a winter coat. He buttoned his blazer and stepped in a soldier's cadence as the marine guard saluted him farewell as he walked into the cold of the evening of a February night and to his limousine. The motorcade cut into the early-evening mess of DC traffic and raced, with lights and sirens blaring, to Joint Base Andrews. The king's aircraft would be airborne within the hour.

The political officers at the US Embassy in Amman remained at their desks well into the night between February 3 and 4.

The email traffic between Jordan and Foggy Bottom was intense. Upstairs, behind Amman Station's fortified doors and soundproof walls, the Arabic-speaking intelligence analysts on staff were busy typing up their lengthy assessments. The State Department and the CIA were both concerned that the release of "Healing the Believers' Chests" would bring mobs to the streets in demonstrations demanding that King Abdullah withdraw Jordan from the anti–Islamic State coalition. Assessments from the Amman street were being reviewed and summarized at Langley as the night shift logged in to their computers. An update would be in the President's Daily Brief next morning.

Fracturing Jordanian resolve for the war was one of Abu Mohammed al-Adnani's main objectives in filming the gory spectacle. It was a gross miscalculation.

Jordanians all over the country did, indeed, take to the streets that night. They carried candles and they draped the nation's flag over their shoulders. Men, women and children representing all of the country's social and economic sectors marched. They paraded in silent prayer in the swanky precincts of Jebel Amman and launched a human vigil in the towns and villages of the desert. People held placards with the picture of the murdered pilot. They carried signs with the hashtag #WeAreMoaz.

Very few Jordanians slept that night. "My stomach turned upside down and I wanted to punch something, someone, I was so angry at what they did to Moaz," a prominent Amman attorney recalled. "Marching with everyone else was one way to show the world that we were united and to show the bastards that they wouldn't win."[150]

The king's aircraft flew up into the winter's night, reaching a cruising altitude for the ten-hour flight back to Amman. It had been a long and grueling day for King Abdullah, but sleep would have to wait—there was still quite a lot of work to be done. He spoke to his generals and his spymasters. He ordered the Royal Hashemite Court to coordinate a condolence visit to Moaz's parents.

The most pressing order of business was the review of paperwork so that Sajida al-Rishawi would be executed before the dawn of the new day. Executions in Jordan were drawn out and complex affairs that involved a slew of bureaucratic hurdles. But the king's orders were carried out immediately and to the letter. The warden of Juweidah Women's Prison was at home eating dinner when the call came from the Royal Hashemite Court. Startled by the urgency of his orders, he nervously put his uniform on and raced back to the facility.

Al-Rishawi was in her solitary cell and fast asleep when the prison's hierarchy woke her up. The warden looked her straight in the eyes as he delivered the news that she would be executed before sunrise. The Iraqi showed no emotion; her expression was blank as if she had been expecting her date with the hangman. She declined a last meal, as well as the ritual Islamic bath to cleanse her body for the afterlife. She was given a set of red fatigues that was worn only by prisoners about to die and was escorted out of her cell under heavy guard.[151] A convoy of armored vehicles and counterterrorist operators were waiting to transport her the sixty miles to the Swaqa prison, the largest in Jordan, south of Amman. Reporters were waiting outside the prison.

The execution was carried out in a small room that recently received a fresh coat of white paint. It was dimly lit and barren; the room's function made it cold and eerie, a foreboding feeling intensified by the harsh desert winds whipping against the windows. Witnesses had been summoned along with several members of the judiciary. A Muslim chaplain recited a brief prayer. Dawn was still two hours away. At roughly 5:00 a.m. Sajida al-Rishawi was escorted up the small flight of stairs that led to an elevated landing. A heavy braided rope hung from the ceiling that was knotted off with a hangman's noose. A lever was connected to a trapdoor that swung open. Several prison officials in their British-inspired uniforms were waiting, as were several plainclothes GID officers; some of them had helped to bring her to justice nearly ten years earlier.

There were no last words, no remorse offered to the victims of the terror cell she belonged to. The noose was tightened around her neck, and at the order of the warden, the executioner pulled the lever that forced the floor cutout to drop. The witnesses looked on as al-Rishawi's neck snapped at the end of the rope; they watched as her body swung from the gallows. A prison doctor confirmed that she was, indeed, dead.

There was another execution in the hours before dawn. Ziad al-Karbouly, a former Iraqi military officer and senior lieutenant in Abu Musab al-Zarqawi's command who was captured in Iraq in May 2006 by GID officers and an assault team from CTB-71, was hanged. He had been arrested for the murder of a Jordanian citizen in Iraq, as well as planned operations in the kingdom. He was sentenced to death and had languished on death row in the Swaqa prison for eight years before he, too, was woken up by the warden in the middle of a cold February night.

The bodies of both executed prisoners were removed into the night in separate ambulances and taken to undisclosed locations for burial. Three other men who were on death row, all members of the Zarqawi network, were also executed before the first rays of light of the new day.

It was already the day after Moaz in Jordan.

10

Kill the Monsters

Revenge is an act of passion; vengeance of justice. Injuries are revenged; crimes are avenged.

—**Samuel Johnson**

AN IMPOSING IRON fence, twenty feet high and painted black with a gold-plated crown in its center, protected the entrance to the Royal Wing from the passenger areas of Jordan's Queen Alia International Airport, located fifteen miles south of the capital. Stern-faced military policemen, wearing rose-red berets and carrying Heckler & Koch submachine guns, ringed the main perimeter and the roadway leading to the section of the airport used by the king and other members of the royal family. Security around the area was no-nonsense. Royal arrivals were usually secretive events known only when a high-speed motorcade consisting of armored vehicles and follow cars equipped with M134 Dillon Aero 7.62 x 51mm Gatling guns that could project three thousand rounds of armor-piercing fire per minute appeared at the site. But on the morning of February 4, 2015, several thousand Jordanians waited in the teasing warmth of the winter sun to greet King Abdullah upon his return from Washington, DC, and the fast departure turned into a slow crawl.

People gathered by the side of the main highway to Amman with English and Arabic banners proclaiming, "We are all Moaz!" Royal Guard and gendarmerie reinforcements sent in for crowd control found it hard to contain the throngs. In an unprecedented display of unity, temperamental Bedouin elders dressed in their traditional robes lined the side of the road in a silent vigil. Jordanians were horrified by what had happened to one of their own. The crime, rather than splinter the country, had helped unify it as never before.

The king returned to the royal palace to a ream full of notes and missed phone calls from world leaders expressing their condolences. His first act upon returning was to address the nation, wearing the same suit and tie that he had on twenty-four hours earlier when he addressed the Senate and spoke with point-blank resolve to President Obama. He donned a red-and-white checkered kaffiyeh and stated, "We have received with all the sorrow, sadness, and anger the news about the pilot, the hero, Moaz al-Kasasbeh. God rest his soul. He was murdered by the terrorist group Da'esh. This cowardly Islamic State group does not resemble our religion in any way."[152]

Later in the day the king met with his general staff and air force commanders to discuss military moves that Jordan would take in response to Moaz's murder. It would be one of the largest Jordanian military operations since the country was founded in 1921.

None of the pilots of No. 1 and No. 2 Squadrons slept much the night after the Islamic State tweet of Moaz's immolation. Some headed out to call on Moaz's widow and his parents; the younger pilots called their own wives and parents to reassure them that they would be all right. "We were angry, and we wanted revenge," said Lieutenant Ra'ed,* an F-16 pilot in No.

---

* A pseudonym to protect his identity.

1 Squadron. "We wanted a mission. A mission to bomb the bastards who killed our friend. A mission to send them all to hell."[153] The king made sure that the RJAF would be assigned a long list of priority Islamic State targets for the next morning. Over twenty targets in Iraq and in Syria were penciled in at Inherent Resolve headquarters. A dozen of the Islamic State barracks and arms depots assigned to the Jordanian Viper drivers were in Mosul; the rest were in Raqqa. US Air Force F-16s and F-22 Raptors would fly escort; KC-135 in-flight tankers would hover safely above the battle space to extend the range and deployment time of the aircraft over the Caliphate. US Air Force and Marine Corps, as well as British and other coalition special operations officers, addressed the issue of close air support and combat search and rescue; the officers were careful about even hinting that there might have been any failures the morning that Moaz was captured. Throughout the night, Jordanian aircrews—including female soldiers—wrote anti–Islamic State slogans on the ordnance to be dropped. One airman wrote the Koranic verse "Let fall on them fragments from the heaven" (34:9) on a GBU MK-82 500-pound bomb destined for Raqqa.[154]

Thirty Jordanian F-16s took off for targets in Iraq and Syria early in the morning of February 5.[155] The aircraft flew together in a tight formation on a northeast trajectory until it split into two: one flight headed north toward Mosul and the other pushed on in the direction of Raqqa. The Jordanian strikes hit over a dozen Islamic State targets—some that were defended by antiaircraft cannons and shoulder-launched surface-to-air missiles. None of the attacking aircraft were hit during the bombing runs; they destroyed all their targets and reportedly killed thirty-five terrorists.*

Instead of returning to Muwaffaq Salti Air Base, the F-16s flew

---

*   Following the air strikes, the Islamic State claimed that US aid worker Kayla Mueller, held for seventeen months by the terrorists, was killed in one of the Jordanian air strikes. Coalition intelligence services believed that Mueller had been killed long before, and that the Islamic State was using the Jordanian bombing run as an attempt to shift the narrative away from the terrible crime that had been posted on Twitter and social media of Moaz's immolation.

over Amman and then headed south, where they performed a victory lap above the village of Aiy and the home of the Kasasbeh family. The Kasasbeh clan set up a huge white mourning tent for the men and a smaller tent just for the women. Women who weren't in the tent, old and young alike, worked in a communal kitchen and over grills to prepare food and sweets for the many villagers, soldiers and even average citizens from all over the country who lined up the twisting dirt path leading to the Kasasbeh home to offer their condolences; the smell of lamb grilling on the fire and fried pastries wafted through the ancient village.

The roar of the F–16s flying their low-altitude afterburner salute to the parents of their fallen comrade energized the long lines wishing to offer their respect. The aircraft would be back over the skies of Mosul and Raqqa the next day—and the days after that—bombing Islamic State targets with greater frequency and ferocity.

King Abdullah arrived in Aiy shortly after the flyover. The Jordanian monarch embraced the Kasasbeh patriarch, and he expressed his sincere condolences on behalf of the entire nation. One of the responsibilities of a commander in chief, especially in a small country, was to personally extend a hug and a salute to those left behind when a warrior was killed in the line of duty. The king and Moaz's father spoke as the two men sat on cushioned chairs at the head of the tent where the male members of the immediate family were gathered. Safi Kasasbeh, Moaz's father, wore a brownish Bedouin robe. His face was coarse and chiseled, carved by the desert winds and by the anguish of watching a group of thugs incinerate his son. The Kasasbeh patriarch turned to the king in a confiding whisper, and said, "I demand that this criminal organization [the Islamic State] be annihilated."[156] Yassin al-Rawashdeh, one of Moaz's distant uncles and a former army officer, went even further. He demanded that the king "kill the monsters!"[157]

King Abdullah's motorcade flowed into GID headquarters shortly before the onset of evening. The convoy of speeding armored vehicles swept past the gates and concrete obstacle blocks of the intelligence service's campus and passed the armored booths and heavy machine gun positions meant to mitigate hostile threats long before they could get anywhere near the spymasters. The king's schedule was packed with planned visits to a long list of army and air force units throughout the country to help bolster morale, but his visit with the spies was purely operational. "The mood at GID headquarters was," an officer in the Counterterrorism Division remembered, "what it must have been in the United States after Pearl Harbor. We were all ready to roll up our sleeves and get busy. We wanted revenge."[158]

Word of the king's visit to the United States raised morale at GID headquarters—especially some rumors concerning his stoic comments to American lawmakers and his requests of military assistance so that the conflict could continue. General Faisal Shobaki met the king at the entrance to the main headquarters. The GID chief brushed off his dark gray suit and straightened his silk tie as the motorcade came to a stop. Both men walked together up the flight of stairs to the second floor, then down a pristine hallway of red carpeting, white walls and framed portraits of the country's kings. GID protection specialists and uniformed members of the Royal Guard escorted the two men into the director's office. The division heads of the service were already waiting. The king sat at the head of the table to convene his war council.

King Abdullah looked around the table at the men who secretly protected Jordan. He made the point to stare at each one of the division chiefs in the eye; he selected his words carefully, and his tone was direct. The king explained that Moaz's murder was a turning point in the war against the khawarij, and it was up to Jordan—and the GID—to deal a crippling blow to the Islamic State. This was also an important opportunity, the king explained, to show the Western intelligence services that

Jordan had the lead in the West's war against fundamentalist Islamic terror. The GID maintained excellent relations with the
espionage arms of NATO and the intelligence services of the
United States and the CIA. The war against the Islamic State
required the very best talents of all involved, the king implored.
No one nation would or could do it alone. The United States and
its intelligence-gathering services with budgets in the hundreds
of billions and satellites had been unable to pinpoint Moaz's location or where the men who held him were hiding. And the
men in the room and the spies they commanded, perhaps the
Middle East's finest HUMINT case officers and analysts, had
also brought pieces of the puzzle to the equation, but not the
complete picture.

The division chiefs were all highly experienced veterans of
the back-alley treachery of the Middle East spy wars. The men
responsible for the Syria Desk, the Iraq Desk, special operations,
internal security and foreign liaisons listened as the call for vengeance was made, wondering what resources they had that could
assist in the endeavor. The CTD, the department's largest division, would bear the brunt of the workload.

Brigadier General Habis al-Hanayneh owned the GID counterterrorism portfolio. His friends, as well as his enemies, called
him Abu Haytham, or the Son of a Lion. Brigadier al-Hanayneh
carried that nom de guerre with passion and pride, and any
overseas section head of a foreign intelligence service wanting
to court his favor addressed him accordingly. Abu Haytham
was shaped like a steel bank vault, and the burly spymaster had
jet-fuel-black hair that covered a medicine-ball-size head; his
mustache was thick and shaped like the kind the Ottoman pa-
shas were fond of. A foreign intelligence officer who worked
with Abu Haytham said that he resembled Omar Sharif if Omar
Sharif were a nightclub bouncer.[159]

Abu Haytham's imposing appearance was just an introduction to what lay beneath. He could be ruthless, a sledgehammer,

showing no mercy to his foes. He was known to have a howit-
zer for a temper, but he was also analytical and calculating. He
carefully pondered every move he made, knowing that counterterrorism was a chess match, where manipulation and patience
were key. He carried a lifetime's worth of scar tissue under his
bespoke suit—it was the collateral damage of sleepless nights
spent on a stakeout, and dozens of late-night shootouts that put
terror cells out of business. Abu Haytham was one of the most
capable officers in the service's history, and he understood the
terrorist mind-set better than most of his contemporaries—in
Jordan and beyond.

The orders were as much for revenge as they were to preserve
national pride. The terrorists calculated that by executing Moaz
in a daylight ceremony, at a time of their choosing, even though
Raqqa was under constant bombardment, the mockery would
openly challenge Jordan and the Inherent Resolve coalition.[160]
The terrorists, the king explained, would have to pay for their
arrogance and cruelty.

The king ordered the GID to target and kill every member of the Islamic State's army that laid their hands on Moaz,
with a preferred priority to go after the Islamic State's *military*
leadership—the men who planned the execution and the senior
committee members who so gleefully approved the decision.
The king told the spy chiefs that he didn't care if the pursuit of
these men took fifty years.[161]

King Abdullah wasn't seeking fanfare or glory, just results. He
wanted the khawarij to disappear into the rubble of history, and
he wanted their deaths to be a warning to anyone with thoughts
of filling their shoes. Inside GID HQ, the ad-hoc name given to
the operation was Guillotine. The nickname name stuck. The
spy chiefs liked it. None of the division chiefs asked any questions about their assignments. Their orders were clear. Each
man knew what was expected of him and the agents in their
command. The men stood up straight and at attention when

the king rose from behind the table. Abu Haytham headed back to his office to convene a meeting with his management team.

Abu Haytham might have been the man in charge of the CTD, but his section chiefs were the men who really ran the show. They were all handpicked and were the bravest and most cunning officers in the entire organization. These midlevel managers—Bedouins, Circassians and Chechens—earned their stripes and a chest full of medals battling al-Qaeda and Zarqawi. They had survived long days that stretched into endless nights in the field and in the office. Most smoked at least a pack a day; they drank an unhealthy amount of coffee. Some carried the physical scars of work inside refugee camps inside Jordan and beyond the nation's borders—from being stabbed or shot. Others had served tours in Iraq and Afghanistan, and they were now very experienced on the Syrian battlefield.

One element of the king's marching orders to what became known as the Moaz Task Force was that the GID work closely with the CIA. The section chiefs were the ones who met with the CIA liaison officers to talk shop, exchange intelligence and barter with one of the region's most valuable commodities: information. One of the section chiefs, the one who usually maintained the links with the Americans, lit a cigarette, took a sip of stale and tepid coffee, and called his contact at the US Embassy.

BOOK THREE

GUILLOTINE

11

Amman Station

The foreigners come out here always to teach, whereas they had much better learn, for, in everything but wits and knowledge, the Arab is generally the better man of the two.

—T. E. Lawrence[162]

THE AMERICAN EMBASSY in Amman is a castle-like fortress sitting on a plateau of prime real estate in the Jordanian capital, in an upscale neighborhood known as Abdoun. The sprawling three-story pink building looks like a mini-Pentagon. The enormous building and the protective compound that surrounds it look very much like they could withstand a firestorm. The embassy was built after Hezbollah's 1983 suicide vehicle bombing of the US Embassy in Beirut and the subsequent release of the Admiral (Ret.) Bobby Ray Inman's *Report of the Secretary of State's Advisory Panel on Overseas Security* that recommended physical measures needed to protect America's overseas diplomatic posts in the age of catastrophic terror. It is reinforced by setback, placing a safe physical distance between any potential truck bombs on the street and blast-resistant windows and doors of the facility in order to lessen the impact of any explosion. To mitigate risk, the two-way vehicular traffic on al-Umawyeen Street in front of the compound entrance is closely monitored: a garrison of heavily armed Jordanian policemen stand outside the main perimeter in full tactical kit; speed bumps and concrete barriers prevent any threatening car or truck from accelerating where it can smash through the main gate and the fortified walls. As the embassy grew in importance, add-ons were built to the existing structure turning the compound into a mini-city running American interests in Jordan and throughout the Middle East.

The ambassador is king in an embassy, or in this case, queen. Ambassador Alice Wells was a powerful presence in the chancery. As *the* representative of the president of the United States in Jordan, Ambassador Wells was responsible for maintaining and advancing the strong bilateral ties that existed between the two countries; when she spoke to King Abdullah, she did so on behalf of President Obama. Ambassador Wells was also in charge of everything that went on inside the embassy and overseeing all activities of personnel assigned to the embassy from the twenty-seven federal agencies that operated outside the United States; when an ambassador convenes a country team meeting, the representatives from the political, economic, management, security, public diplomacy as well as other agencies sit at the table.[163] The embassy also fields more secretive offices, including military attachés and representatives from the Department of Homeland Security and the Department of Justice; even departments such as the Transportation Security Administration have a desk. The CIA office, known as a station, was sequestered in a well-protected corner of the embassy out-of-bounds to anyone who didn't have direct business with the Agency.

In the CIA's organizational chart, Amman Station fell under the command of the Near East Division—one of the Agency's regional commands.* The Near East Division was one of the largest and most important in the Agency, covering some of the most problematic countries in America's war on terror. Iraq and Lebanon were part of the division, as were Saudi Arabia, Qatar and Israel. Amman Station was considered one of the most ambitious postings for CIA personnel, as the location, a hub of Middle Eastern intrigue, enabled a case officer to be in touch with Jordanians, Syrians, Palestinians and many other persons of interest who passed through the Hashemite Kingdom.

Jordan was a front line in the war against the Islamic State, and "Amman is very much like Casablanca in the Humphrey Bogart film," a retired employee of the US Diplomatic Security Service once commented.[164] Amman was a coveted post and there was fierce competition at Langley to be assigned to the station. The country was stable from within and friendly toward westerners. Embassy—and station—personnel could bring their families with them, and although security concerns existed, the country was a place where there were Starbucks, top-flight restaurants and a true sense of history that didn't exist in many other frontline countries.

Amman Station was a paradox. A case officer could float with his family at the Dead Sea, enjoying mineral mud and massages in the morning and be at the frontier with Syria meeting a source against a backdrop of civil war for afternoon tea. Amman was a common travel connection for the personnel. Tel Aviv was an hour's drive away. Beirut was eighty-five minutes away by plane; the flight to Baghdad was ten minutes longer. "The city, hell the country, was a crossroad," a British diplomat commented. "It was a launching pad for deals inside the kingdom and aspirations beyond the country's borders."[165]

---

* On October 15, 2015, CIA director John Brennan reorganized much of the CIA, and the traditional divisions were reorganized into what became known as Mission Centers. The Mission Centers included Africa; Counterintelligence; Counterterrorism; East Asia and the Pacific; Europe and Eurasia; Global Issues; Near East; South and Central Asia; Weapons of Mass Destruction and Proliferation; and Western Hemisphere.

Most of the men and women working the station had been preordained for work in the Middle East at the time of their hiring or soon after at Camp Peary, known in the Beltway vernacular as "the Farm," the legendary CIA training center in Williamsburg, Virginia. CIA talent spotters and Farm instructors would earmark candidates with Middle Eastern ethnicity, language skills or a burning desire for future work in the region, and after they completed the tradecraft training all case officers must complete, then their instruction, be it language school or something more specialized, would be designed according to where they would be sent. At an advanced stage of the training, a former case officer named Ted* explained, "The heads of the various regional divisions would come down to see the trainees and to sell them on working in this country or that. The future officers could have a preference and part of the process is for them to bid on open slots and opportunities. The division chiefs always wanted to make sure that the best man or woman went to where they were needed and could be most useful."[166]

Once training was completed, the newly minted case officers destined for overseas work selected their posts according to personal preference. Those who liked smaller "shops" where they'd have more latitude and adventure could select a small embassy. Others, with special skills or with families, would select a larger embassy that had proper living accommodations and American-accredited schools for their kids; in some locations where no such facilities existed, the children were sent to a boarding school in a third country. A complete in-house embassy workforce was involved in finding homes and cars and other amenities for the newly arriving officers. Living conditions were important. Tours lasted two or three years and there were often opportunities to extend. "An unhappy home life threatened an officer's marriage and their effectiveness," Ted reflected, "and it was a reason—that and alcohol—why people stationed over-

---

* A pseudonym.

seas had such a high divorce rate or requested to be reassigned stateside."[167] Perhaps most important, Amman was a launching pad for promotion. The Near East Division was where the post-9/11 superstars were made.

American intelligence officers had been at war for thirteen long and exhausting years by the time the CIA took a central role in the Inherent Resolve coalition to destroy the Islamic State. Nearly everyone in the division—and many at Amman Station—had spent a few years of their lives in Baghdad, at the largest US embassy and CIA station in the world. Others had earned their stripes working in Yemen, Libya and Egypt. Others, of course, had carried out campaigns in Afghanistan and Pakistan. The Agency's role in the global war on terror was both controversial and successful: infamous for black sites, waterboarding and Abu Ghraib, yet countless plots disrupted and thwarted.[168] There were some stinging defeats, such as the al-Qaeda triple agent who blew himself up at Camp Chapman, the CIA forward operating base in Khost, Afghanistan, on December 30, 2009, which resulted in nine CIA dead.* Yet, there had also been some incredible successes.

The CIA's relentless pursuit of Osama bin Laden earned its well-deserved dividends on May 1, 2011. The intelligence-gathering and detective work was conducted around the Middle East and across the world. There were also numerous high-profile terrorist targets—al-Qaeda and others—who were taken out by drone, by JSOC or by the Agency itself.

During the war on terror, the CIA of old had forever been

---

* Among those killed in the attack were Darren LaBonte, an Amman Station case officer, and Sharif Ali bin Zeid, a GID officer and a cousin of King Abdullah. The GID had warned the CIA that in their view the "triple agent" they thought would lead them to Ayman al-Zawahiri and, ultimately, Osama bin Laden had been flipped and was working for al-Qaeda and advised that the source be handled with great caution; the GID warned that the man, Dr. Humam Khalil Abu-Mulal al-Balawi, should be searched some two hundred meters before entering the American base and that he be strip-searched to make sure he wasn't concealing a suicide vest or improvised explosive device. Tragically, the GID warnings were ignored, as was Agency protective protocol.

altered from a classic espionage force into a dynamic proactive tactical spear of American interests and policy. CIA stations in the Near East Division had become wartime frontline outposts. Some of the operations that commenced out of Amman Station rippled across the Middle East and beyond. It was, in the words of one former security officer, an "espionage hub."[169]

And like any center of covert activity where only the chief of station knew the objectives and purposes of all those using the facility as a base, the personnel inside the office represented many of the CIA's directorates and units. Communications specialists and translators who understood a dozen dialects of colloquial Arabic worked around the clock, as did technicians who could fix the gadgets, radios and computers that kept the station in touch with officers in the field and with headquarters. The station staffed analysts who had an encyclopedic knowledge of the region and the terrorist armies that America fought. And, of course, there the station hosted a steady stream of personnel on temporary duty, or TDY, in country. Those on TDY included veteran spy hands who were academics and analysts, and those who were James Bond–like shooters from the Special Activities Division, known as by the sinister acronym of SAD.

Protective officers from the Agency's security directorate were also posted to the station. These tactical specialists were responsible for safeguarding station personnel who had high-risk meetings with assets and sources in the field. Years ago, CIA case officers, such as the legendary spy turned spy-novelist Charles McCarry, never carried a gun in the field.[170] The work was dangerous, but a case officer's success or failure inside a darkened safe house for a meeting with a squirrely asset depended on trust and secrecy—not a Brazilian-made Taurus 9mm semiautomatic pistol with the serial number filed off. But the rules of the espionage trade changed drastically in the age of suicide terrorism when the players in the shadows followed no rules other than expediency. The Agency's view of security for its officers in the field changed dramatically following the kidnapping and murder of CIA station chief William F. Buckley in Beirut in 1984, when the Agency's number one man in a hostile country was no longer free to live alone inside a hostile neighborhood unprotected by overt and covert security measures. When the global war on terror came around, those measures involved the work of contractors.

The CIA employed a classified number of former US special operations soldiers as well-paid guns for hire who augmented security and other tactical requirements at stations throughout the global war on terror. The contractors were known as the global response staff, or GRS. The evolution of the GRS as an integral part of CIA operations worldwide was a byproduct of the escalation of the Agency's paramilitary mission. GRSers worked on short-term employment deals offering recently retired personnel with unique skills and top secret or beyond security clearances with the opportunity to earn a substantial payday working as an added layer of security for proactive intelligence work in some of the most dangerous precincts in the world.[171] GRS personnel were among those killed at Camp Chapman in Khost in 2009; two GRS operators, former SEALs Glenn Doherty and Tyrone Woods, were killed defending the CIA's covert annex in Benghazi on September 11, 2012. Despite its dangers, for many former soldiers, it was a good gig. Many working in the GRS did the same jobs they did before as part of JSOC or US-SOCOM, but now as Agency contractors they were paid infinitely more for their risk and troubles. There were numerous and often nameless contractors coming in and out of Amman Station at any given time.

There were two types of CIA officer at an embassy—the declared and the undeclared. Declared personnel were known to both the Jordanians and to the Americans inside the chancery to belong to the Central Intelligence Agency. Undeclared officers worked undercover, assigned to another section inside the embassy while their real mission was clandestine activities and intelligence gathering. These officers often also operated in alias when meeting with their assets. Although the CIA and the GID maintained excellent close-knit operational and intelligence sharing ties, spies were, after all, spies, and their mission was espionage. A popular saying inside the intelligence world was that there were no friendly espionage services.

Undeclared case officers dressed the same as their embassy colleagues. Male officers typically wore jackets and ties to work. Women typically wore business-casual skirts or pants with a jacket. It was necessary for station officers to blend in and assimilate into the mundanity of the embassy so that those in the employ of the Central Intelligence Agency wouldn't be tagged as spies. Officers who came to work and then went out in the field usually had to change their clothes in the field, or at home, or at one of the safe houses in the city that the station rented for meetings. It was important for those in the station to not stick out as being CIA.

There were always meetings going on inside the embassy and, specifically, inside Amman Station. There were case-oriented meetings, staff meetings, station meetings, security meetings, secure link with Langley meetings and so forth—the list was endless. The chief of station always sat at the head of the conference table in the Sensitive Compartmented Information Facility, or the SCIF—while the deputies sat close by. The SCIF was a room that was forced-entry resistant and soundproof so that eavesdropping from the outside was impossible. Amman Station, being in one of the larger and more spacious embassies, had several conference rooms designated as SCIFs. The rooms were Spartan, though complete with the typical Agency amenities—a flag, portraits of President Obama and Vice-President Biden, and some government-acquired artwork that looked like it was purchased in a hotel estate sale.

Jake* was the chief of station—the Agency's most powerful man in country. He was a soldier and had spent a career as an operator inside the US Special Operations Command. Ruggedly handsome and chiseled, he looked, as a former Jordanian intelligence officer commented, like someone a film director would cast to play a middle-aged man fighting a young man's war. As the man in charge of the sprawling intelligence center, Jake was responsible for corralling all the different officers, analysts, deep-cover operatives and TDY shooters into an effective mission-oriented intelligence-gathering machine. Chiefs of station were relatively high on the Agency totem pole—usually a GS-15, which was equivalent to a colonel in the military. Most chiefs of station had at least twenty years inside the Agency and had survived dangerous missions, wars and the political backstabbing that was ever present at headquarters. The chief of station was part manager, part commander, part bean counter, part social worker and part den mother; there were many different personalities inside the insular world of the intelligence section, and all complaints, personal issues and mission obstacles made their way to the officer in charge.

A good chief of station was able to keep his officers safe from hostile forces and from intrusive and obstinate ambassadors and deputy chiefs of mission; and to avoid rebukes from headquarters or threats from the ambassador—the person who was truly in charge of everything that went on inside the embassy and the country. There were two types of ambassadors—career foreign service officers who dedicated their lives to advancing the message of America and all it stood for, and political appointees who were often very rich people, sometimes of dubious intellectual faculties, who had donated a king's ransom to the man who won the presidency. Sometimes ambassadors went native and thought of themselves as more Arab than American; other

---

* A pseudonym.

times, they thought of themselves as James Bond or George Smiley and were so intrusive into the work of the station that the fight was escalated back to Washington to be carried out between the CIA director and the secretary of state.

The chief of station was also the commander of a vast bureaucratic industry. Paperwork, the mind-numbing endless checklist of government redundancy, was the bane of an officer's existence. There were reports of all kinds to file: progress reports, expenditure reports and agent debriefs. "Every time a case officer met with an asset, a report had to be generated about where the meeting was held and why it was held and what was accomplished," Ted remembered. "If the meeting produced intelligence information, one or more intelligence reports were written. Each intelligence report required a separate report explaining how and from whom the intelligence information was obtained. If no intelligence information worthy of dissemination in an intelligence report was obtained, often an additional report concerning the information collected which was useful, but not intel report worthy. There was a report that had to be filed explaining expenditures, if any. Some of the reports were in-house, for station eyes only. Other reports had to be sent back to headquarters, copied on the email chain. The chief of station had to approve all communications traffic between the office and headquarters. 'I spent more time writing reports than I did in the field gathering intelligence' was a popular gripe among Station personnel.

"Officers who were undeclared [operating in true names and assigned to other departments to shield their Agency affiliation] were worse off," Ted explained. "In addition to performing the duties for their Agency jobs, they had to perform the duties of their cover jobs as well, such as having meetings with their host country counterparts, writing reports, and attending various embassy meetings with their non-agency colleagues."[172] Everything a member of the station prepared to be emailed to headquarters was reviewed by the chief of station.

Mobile-phone communications and apps that enabled encrypted chats between individuals and groups were both a curse and a blessing for a chief of station's often-complicated need to control the back-and-forth of information. Smartphones and throwaway relic cell phone models allowed officers to remain in close contact with their assets and sources in the field and to do so anonymously. Iraqi, Syrian, Palestinian and, of course, Jordanian SIM cards were inexpensive and easy to buy in Amman or north near the refugee camps. Mobile phones were ridiculously cheap, and many in the station carried two or three. But nonissue mobile phones were a security risk—they were easy to monitor, and if one was lost, it could reveal compromising data if in the wrong hands.

Jake's most important job was his role as the Agency's intelligence conduit to the GID director.

The man who commanded the GID's international liaison unit that interacted with all other friendly—and not so friendly—intelligence services around the world was a straightforward career intelligence officer who dealt with the other Western and Arab espionage agencies that were in routine contact with the GID. He was low-key and calculating and known in the service for plotting long-range and time-consuming operations that involved cunning. He looked unassuming and scholarly. He was considered one of the GID's most analytical minds, and he was a fine choice to be the link to the rest of the world's espionage service fraternity.

The CIA had a different relationship with the GID. Jake maintained direct links to the GID director and routinely met with him to discuss the pressing matters of the day, as well as special joint operations, such as the hunt for the men responsible for Moaz's murder.

Jake's deputy, a veteran intelligence officer named Josephine,* maintained direct communications with the GID's deputy commander. It was an outside-the-box way of maintaining an intelligence partnership that protected highly complex geopolitical considerations, but it worked, and that was all that mattered. The number one and number two in station spoke of not only the day-to-day details of intelligence sharing and cooperation but also the policies that these operations influenced. Syria and everything it touched was a source of concern.

They now worked together to avenge Moaz.

The US Embassy—including the station—followed a Jordanian work schedule and was open from Sunday to Thursday. Sunday mornings were reserved for the weekly in-country meeting at Amman Station. The chief of station convened the gathering at an hour of his or her choosing. The chief of station always sat at the head of the conference table in the SCIF, and most of the case officers and analysts assigned to Jordan needed a good excuse not to be there. The chief of station, and others who sat behind a desk, always wore jackets, shirts and ties; the women wore business suits.

On Sunday, February 8, 2015, Jake convened the usual start of the week's in-country meeting. The deputies, the officers and those from the clandestine services and the more tactical side of the company made their way into the SCIF, juggling notebooks and snacks as they did at the start of every work week.

Once the meeting ended, Jake gathered his folders and his metal thermos cup full of black coffee and headed to his office. He closed his door and called the palace to see if he could meet with the king. Hours earlier he had received instructions to assist the GID in the hunt for the Moaz killers.

---

* A pseudonym.

Royal Protocol, the men who handled King Abdullah's daily affairs, had already reshuffled the monarch's busy schedule. GID director Faisal Shobaki would be waiting, as well.

12

The Kill List

He that spies is the one that kills.

—Irish proverb

BEHIND CLOSED DOORS, in the operational workspace and internal communications of GID existence, each international espionage service was officially referred to by a code. Unofficially, of course, there were nicknames. The CIA and the Americans were known as the "Friends" and also the "Big Boys." The affectionate CIA nickname for the GID was the "Jords."

The CIA and the GID were partners with shared enemies and short- and long-term objectives. But they were also competitors for the most valued product in the region: information. Their relationship had nowhere near the level of trust and intimacy that the Americans shared with the "Five Eyes" intelligence alliance. The domestic and overseas intelligence services of the United States, Great Britain, Canada, Australia and New Zealand constituted the Five Eyes arrangement, and these nations worked alongside one another without hesitation or suspicion. The sharing of intelligence was intimate and un-interrupted; they even shared secure access to classified lines of communications, such as SIPRNet.[173] The phrase "There are no friendly intelligence services" was hammered into the heads of new CIA trainees and federal law enforcement officers un-dergoing their basic agent courses; the words were written in capital letters on blackboards in just about every counterespio-nage classroom.[174] There was always concern that any intelli-gence shared could ultimately be given or bartered to another agency or divulged accidentally or used maliciously.* Intelli-gence, after all, was currency.

The day-to-day interactions between friendly intelligence services were transactional, and the currency that was bartered, sold and held in savings accounts was information. Intelligence was traded like stocks on Wall Street; great intelligence earned interest and bought respect and favors for a later day. "A friendly service bearing gifts always wanted something in return," a retired Middle Eastern case officer and section chief who re-quested anonymity explained, "and when you have to recipro-cate and give something back, make sure it's sanitized and won't come back to bite you down the road."[175] Suspicion and appre-

---

* An example of which occurred in May 2017, when there was great furor in in-telligence headquarters around the world when US President Donald J. Trump allegedly handed sensitive information concerning a human asset inside the Caliphate to the Russian foreign minister and the Russian ambassador to the United States, whom he hosted in the Oval Office. The top secret dossier de-tailed an Islamic State plot to use sophisticated laptop computer bombs to take down civilian airliners was, reportedly, provided by a deep-plant spy working on behalf of Israel's Mossad and then disseminated to the CIA. By provid-ing the Russian officials with the highly classified report, it was claimed that President Trump's action not only violated the compartmentalization of the enormously sensitive information and betrayed the trust of an allied agency, but it also exposed the identity of the human source and exposed him and his family to personal harm. There were even reports that the president's actions resulted in a temporary freeze on the bilateral exchange of intelligence be-tween the Israeli and American intelligence communities. But other reports, for example the May 18, 2017, report by Ali Younes for Al Jazeera ("Jordanian Spies Provided ISIL Bomb Intel: Officials"), hint that the actual source of the intelligence was a Jordanian asset, and that Israel did not have any HUMINT sources deep inside the Caliphate. Still, such carelessness displays the inherent weakness in sharing sensitive source data.

hension, even among friends, was routine and part of the social contract of the spy game. When the GID met with their CIA liaison counterparts from Amman Station, officers from both sides knew exactly what and what not to say.

On a cold, rainy afternoon in the Jordanian capital, a week after the Moaz immolation video was transmitted around the world, Jake and Josephine had already met with the king and with Faisal Shobaki to confirm priorities. Winning the war against the Caliphate was the geostrategic imperative; revenge for Moaz was a national must for Jordan, and the two senior CIA officers were reminded of this imperative over and over again. The details—the operational machinations—would be left to the liaison and special ops representatives from both sides.

The American visitors were escorted through security checks in the outer perimeter and led to the CTD complex, where they were brought to a conference room reserved for interservice liaison. The hosts met their guests inside the room. Sit downs never began before an attendant brought in Bedouin coffee with cardamom poured into a small cup followed by a glass of piping hot mint tea. The door was closed once the traditional and time-honored hospitality was completed. Sometimes the GID liked to take the Friends to landmarks such as the remote Crusader castle at Shobak, built in the southern desert in AD 1115 by King Baldwin I, or to hilltop vistas that overlooked the Dead Sea and Israel and where the lights of Jerusalem could be seen at night, or even to one of Amman's famed all-night hummus eateries, like downtown's legendary Hashem's, but that was the extent of the fraternization. "There were no invites for a home-cooked meal, or any other family events," a retired GID officer explained. "The officers from both sides could have genuinely liked one another, they could have had a million things in common, but the relationship was always professional, not personal."[176] The GID hosted all official exchanges and the formal ones were always held at the GID headquarters campus in Wadi Seer. Discussions of the Guillotine manhunt were conducted inside headquarters.

The American visitors were escorted through security checks in the outer perimeter and led to the CTD complex where they were brought to a conference room reserved for interservice liaison.

A GID officer named Bashir chaired the initial meeting. Bashir wasn't his real name, of course—operational pseudonyms and cover stories were par for the course for those sitting on either side of the conference table. Bashir didn't reveal his rank or his precise job title, though he routinely interacted with the Agency when it came to matters involving counterterrorism. Bashir was a twenty-year veteran of the GID and had spent much of his career in the CTD, rising in rank because of some great work rendezvousing for clandestine meets at ungodly hours with nothing to protect him except his instincts, his courage and a pistol. Bashir was short and as solid as a linebacker, with a no-nonsense physique of muscle that hinted at years earlier when he might have been quite the go-getter in the field. A slight scar cut across his chin, he walked with a limp that he fought hard to hide, and along with his close-cropped haircut he sported the military mustache that was a reminder he had served in the army before becoming an intelligence officer. The GID often favored former soldiers.

Bashir did his best to welcome his guests—Jordanians were renowned for their hospitality—but he struggled to produce a smile. A grin was a sign of weakness. Niceties were the responsibility of the other men in the room.

Bashir's English was rudimentary and rough. Although his guests spoke Camp Peary–taught Arabic, another GID officer, named Suleiman,* was on hand to translate every sentence into flawless English with a hint of a Southern twang. There were

---

* A pseudonym.

quite a few GID officers who, while in the military, learned their English in North Carolina at Fort Bragg, in Georgia at Fort Benning or in Kentucky at Fort Campbell.

Bedouin coffee was served in small white porcelain cups. The coffee was to be consumed in front of the server, and if the drinker shook the cup, it meant he had had enough. Usually one cup sufficed. Light and greenish in color, the aromatic Bedouin coffee laced with cardamom was nothing short of rocket fuel. The coffee was followed by sweet mint tea in small glass cups served with an offering of Arabic sweets from a colorful tin tray. A dozen bottles of mineral water were on the table. Business could not commence until this Bedouin hospitality ritual was complete.

Bashir welcomed his guests and explained that although Combined Joint Task Force–Operation Inherent Resolve placed an emphasis on targeting high-level Islamic State officials, the most pressing mission entrusted to the GID was to locate the men responsible for Moaz al-Kasasbeh's capture and murder and make sure they were destroyed. How each of the targeted individuals would die was unimportant at this stage.

Bashir went to great lengths to explain that the GID had a task force that had been working on identifying the men held responsible for Moaz's death from the day he was captured. The intelligence that identified these specific men as being commanders or facilitators of the grisly murder drew upon sources that were deep inside enemy territory, and were highly reliable.

Several other GID officers entered the room and sat at the polished conference table. Most looked as if they were well into their twentieth year of work; nicotine stained their fingers and there were coffee stains on the paperwork they carried. All looked like they had been deprived of sleep for too many nights. Bashir didn't introduce his colleagues.

One of the GID officers in casual dress held five white plastic folders. There were only a few pages in the files with only a handful of printed mug shots or other photos found in open-source archives that depicted the names and images of each of the most wanted men. Bashir explained that assets and sources inside Syria had been able to unequivocally identify each man on the list as being the ones who decided on Moaz al-Kasasbeh's cinematic murder. Bashir didn't elaborate more on what or who those sources were. The Amman Station liaison officers knew better than to ask.

Abu Bilal al-Tunisi's name was not on the official wanted list, but his name was the first one discussed. Although al-Tunisi claimed Moaz, then beat and humiliated him for sport, he was of relatively low rank. The chubby and sadistic Tunisian didn't warrant a bilateral multiagency task force hunting him. His death would be one of opportunity.

Abu Mohammed al-Adnani was the highest-ranking Islamic State official marked for death. The GID assessed that killing the sadistic social media mastermind was critical to defeating the Islamic State altogether. He was known in counterterrorist circles simply as Number One.

In his role as the commando leader, spy hunter and marketing genius behind the Islamic State's murderous films and online recruitment videos, al-Adnani proved himself to be a brilliant asymmetrical warfare tactician. He had a highly trained and fanatical security detachment permanently assigned to him. He was wily with uncanny instincts, the GID assessed, and moved around constantly to conceal himself from coalition drones and satellite coverage. Al-Adnani, the GID reports indicated, was psychopathically paranoid. He trusted no one and suspected everyone. He was known to accuse even those Islamic State fighters who were close to him of being on the payroll of the CIA or the Mossad; the Russian FSB and GRU; Hezbollah and the Iranian Ministry of Intelligence Services, the MOIS. Enemies were everywhere, and al-Adnani didn't require that suspected traitors be proven guilty beyond a shadow of a doubt. Men merely suspected of a dual allegiance were tortured and executed by his fanatical secret police. There were those in the intelligence community that believed that the torture he endured as a political prisoner inside Syria's prison system turned him into a pure psychopath.

Number two on the list was Abu Omar al-Shishani, the Islamic State's war minister. Bombastically arrogant and known to be someone who relished in the savage disposal of the Caliphate's enemies, al-Shishani personally supervised Moaz's time in captivity and then helped supervise the production of his execution.

Abu Omar al-Shishani was born Tarkhan Tayumurazovich Batirashvili on January 11, 1986, to a Georgian Christian father. His mother was a Kist, an ethnically Georgian Sunni Muslim. Batirashvili grew up in the small village of Birkiani in the remote Pankisi Gorge near the Chechen border, a mountainous and lawless region known for fundamentalist clerics and men who lived by their trigger fingers. Much to his father Timur's chagrin, Tarkhan and his two brothers converted to Islam.[177] The brothers were obsessed with the fight against the Russians as part of a larger Islamic struggle.

As a young man, Tarkhan saw Islamic fighters from around the world passing through his region on their way to fight Russians in the Second Chechen War. The men with long beards and flowing robes marched through the inhospitable hills to face certain death, but their heroics and faith inspired Tarkhan. According to legend, the young redhead even joined the men from Arabia on missions against Russian forces.[178]

Tarkhan Batirashvili was conscripted into the Georgian military in 2004 and served as an NCO. He was an operator in an elite intelligence-gathering reconnaissance unit and received advanced counterinsurgency combat training from US Army Special Forces teams sent to Georgia to train their fledgling special operations units.

When the Russians invaded Georgia in 2008, Tarkhan fought in some of the bloodiest battles, earning a reputation as a courageous and daring soldier under fire. According to his father, Tarkhan found a home in the army—a place where he belonged and could build a career, a life and a family—but he was stricken with tuberculosis and forced out of the service.[179] He grew disillusioned with civilian life and was arrested in 2010 for stockpiling weapons that were destined, it is believed, for Islamic groups in Georgia; he had grown a long reddish beard and shaved his mustache like other pious militants and was a fixture at some of the Salafist mosques near the Chechen border. Tarkhan was sentenced to three years behind bars for his crimes but was freed after only sixteen months; the Georgian Intelligence Service thought it best to get him out of the country before he became a homegrown terror commander. Tarkhan was given a passport that was valid for ten years and bus fare to the Turkish frontier. The Georgian government then alerted friendly intelligence services of the passport number, 09AL14455,[180] and issued a warning that Tarkhan could be a person of interest.

It isn't known if Turkish intelligence agents surveilled Tarkhan as he traveled through the country toward Syria. The country was like Afghanistan a generation earlier, a magnet for jihadists around the world who were unemployed, lost, and looking for a war to fight and a cause to justify the violence. Tarkhan crossed the border into Syria in late 2012 and found a group of like-minded men from the Caucasus looking for combat. He formed the Jaish al-Muhajireen wal-Ansar, a guerrilla group of Arabic-speaking fighters from the North Caucasus. He adopted the nom de guerre of Abu Omar al-Shishani: Father of Omar the Chechen.

The men from the Caucasus distinguished themselves in the fight against pro-Assad forces and their Iranian and Hezbollah allies. It didn't take long for the Georgian-born militant to find an army in need of combat veterans that had undergone extensive Green Beret counterinsurgency training. By

2013 al-Shishani swore his allegiance to the Islamic State and, with his tumbleweed-red beard, became the military face of al-Baghdadi's growing army, serving as its war minister and a key voice on its decision-making council. In the field al-Shishani never wore a face-concealing balaclava. He was the Islamic State's General Patton—arrogant, bombastic, capable and a motivator of men on the battlefield. Photographers from the Raqqa Media Center went to great lengths to capture the Georgian in the field surrounded by masked foreign fighters. Images of al-Shishani helped to recruit thousands of young men from Chechnya, Georgia, Uzbekistan and Tajikistan into the ranks of the Islamic State.

Abu Mundhir Omar Mahdi Zeidan was the third name on the list. The GID was intimately familiar with Zeidan—his file in headquarters was as thick as a phone book. Zeidan grew up in a Palestinian refugee camp in the rough-and-tumble city of Zarqa. Pious yet radical, he immersed himself in religious study and ultimately earned a bachelor's degree in Islamic Law. He became one of the most prominent thinkers and impassioned clerics in the global Salafi jihadist community. Zeidan's zeal and support of violence placed him on the radar of the Jordanian security services. He was ultimately arrested and convicted of sedition and terrorist crimes and sentenced to prison along with Abu Musab al-Zarqawi. Zeidan was pardoned in 1999 in the same amnesty that released Zarqawi onto the world. He ventured east toward Pakistan and the global jihad of al-Qaeda and other fundamentalist groups. Zeidan crisscrossed the Middle East to preach radical and Salafist sermons in mosques and in online videos. He was seen quite a lot in Yemen in support of al-Qaeda in the Arabian Peninsula, even eulogizing a fighter killed by a US drone strike.[181]

Zeidan surfaced in Syria in 2013 and pledged his allegiance to Abu Bakr al-Baghdadi. He became a valued cleric in the Caliphate, and welcomed into al-Baghdadi's inner circle, emerging as a prominent advisor on sharia law. Zeidan was charismatic and convinced al-Baghdadi to appoint him as the Islamic State's de facto education minister. Zeidan approved the curriculum that would be taught to all boys inside the Caliphate; he wrote textbooks that were designed to teach the next generation of fighters how to be absorbed in faith and consumed in violence.

Zeidan was ultimately rewarded for his value and appointed as sharia judge and governor of the judicial council—a legal body of scholars and firebrands that was akin to a supreme court.[182] Zeidan believed that Islam needed to clean its own house before it could fight the Shiites, the West and Israel. "Our people in Libya, Afghanistan or Iraq are fighting disbelievers amongst their own just like the Prophet fought his people," Zeidan said. "The Prophet didn't start with fighting the Romans or the Persians, but he started by fighting his people and made the criterion in that to be this religion."[183]

Because Zeidan was a Jordanian, al-Baghdadi hoped that he could lead the march south into the Hashemite Kingdom. Zeidan was eager to please the caliph. He never forgave the country of his birth for placing him behind bars and for cooperating so closely with the United States in the war on terror. And he never forgave the United States for killing his brother Mahmoud in a drone strike in Pakistan: Mahmoud was one of al-Qaeda's top clerics and the personal bodyguard for Mustafa Abu Yazid, al-Qaeda's number three and its financier. He was killed in retaliation for the Camp Chapman murders.

Zeidan crafted and then issued the fatwa calling for Moaz al-Kasasbeh's murder. When some in the Islamic State's ruling council called for mercy to be shown to the Jordanian pilot, Zeidan gave fiery sermons demanding his death.

The fourth name on the list was that of Abdul-Hai al-Muhajir, an Iraqi veteran of Zarqawi's network who was the wali of Damascus. Short tempered and with a penchant for vengeance, al-Muhajir was a member of the Islamic State council who argued passionately that Moaz needed to be killed. Al-Muhajir also urged al-Adnani and al-Baghdadi to open a military front against Jordan and to make King Abdullah pay for his part in the Inherent Resolve coalition. Al-Muhajir was considered influential and a member of the next generation of Caliphate leaders who would assume command of the organization should something happen to al-Baghdadi. His ambition was dangerous—especially his calls to open a southern front and to expand the war. As the wali, he authorized the diversion of vital resources—foreign fighters and combat vehicles—away from the frontline battles against Assad's forces to reinforce the border area in the south near the Jordanian frontier.

Abu Khattab al-Rawi was the final name added by the GID's Moaz Task Force. An Iraqi and a fraternity member of the Zarqawi campaign against American-led forces, al-Rawi was a capable operations commander considered to be a rising star in the Islamic State's hierarchy. Al-Rawi was Abu Mohammed al-Adnani's deputy for external operations, as well as the man responsible for funneling fighters along the corridor to the desert crossroads bridging Syria and Anbar Province in Iraq. "Every plan, every campaign, and every whim that al-Adnani had was ultimately facilitated by al-Rawi," one of the unnamed GID officers in the room broke his silence to explain. He added, "He was his [al-Adnani's] right-hand man, his executive officer, and the organizational chieftain that made things happen."[184]

The meeting that February afternoon was to coordinate the intelligence-gathering aspect of the hunt. The men in the room weren't from the special operations side of the shop. There would be no special operations foray into the Caliphate to hunt the men on the list. It was too dangerous. Syria and Iraq were not like Western Europe years earlier when Israeli agents roamed the streets of London, Paris and Rome, hunting down the Black September terrorists who masterminded and facilitated the 1972 Munich Olympics massacre. The Caliphate wasn't Gibraltar either where in 1988's Operation Flavius, operators from Britain's 22 Special Air Service preemptively stopped an Irish Republican Army bomb plot by gunning down three terrorists as they sat in a car. The Islamic State was fraught with danger and indiscriminate murder. Men and women merely suspected of betraying the Caliphate were murdered in cold blood, often without a trial. Men like al-Adnani and al-Shishani were insulated from danger by a phalanx of bodyguards. They were protected like heads of state. Getting into Syria or Iraq didn't necessarily guarantee getting out. The intelligence personnel were too valuable—and knew too many secrets—to risk. As one GID officer commented, "We were told to write history, not a Hollywood screenplay."[185]

It didn't matter if the targets were strangled slowly with a piano-wire garotte or obliterated by Hellfire missiles fired from a drone. But there were rules. The coalition's targeting process was efficient and effective. The intelligence gathered was reviewed internally, analyzed, scrubbed for any revealing and telltale tidbits that might reveal a source's identity, and then shared with the intelligence services or the coalition operations desk. The strict Inherent Resolve use-of-force guidelines were to be followed in the pursuit of the names on the list. Civilian casualties were to be avoided at all costs.

The meeting lasted another fifteen minutes. Although the Moaz list was the main topic of conversation, there were always other items important to both sides that had to be attended to in the face-to-face sit-downs. Intelligence was a currency in alliances, and the spies spent theirs wisely when interacting with their counterparts. The more volume and better quality of the intelligence—raw or refined—that one side of the table could offer to the other, the more favors were exchanged back and forth on other matters. It was all very Byzantine and incredibly effective.

An attendant brought cookies and sweet Middle Eastern delicacies to indicate that the briefing portion of the meeting had come to an end. The CIA officers remained in their chairs and discussed scheduling their next get-together.

The two CIA liaison officers returned to Amman Station and briefed Jake. The names on the list came as little surprise to the chief of station or anyone that followed the classified comms going back and forth between Langley and the Middle East.

The two officers—wired tightly from the caffeine and sugar overload of coffee, tea and honey-drizzled Jordanian sweets—sat at their desks to write up reports and memorialize the discussions with the GID. The five names on the list, plus al-Tunisi, were relayed in reports that would be transmitted before nightfall. Copies would be sent to Langley, a variety of other CIA stations in the area, CENTCOM, JSOC and USSOCOM. Liaisons to the Five Eyes services would also receive a copy.

The information was hardly breaking news. The names of the Islamic State's political, religious and military commanders were well-known to the generals, intelligence commanders and operations officers from the fifteen nations headquartered at Camp Arifjan in Kuwait who were responsible for prosecuting the war against al-Baghdadi's Caliphate as part of the Inherent Resolve coalition. From the onset of military sorties flown against terrorist targets inside Syria and Iraq, the coalition targeted the men at the top who counted the money, issued the fatwas, directed the military advance across the Levant and spread the fanatic ideology across the world. Abu Mohammed al-Adnani and Abu Omar al-Shishani were highest-priority targets even before Christmas Eve 2014 when Moaz al-Kasasbeh was seized near Raqqa. But the kill list told the men in charge of Combined Joint Task Force–Operation Inherent Resolve that killing the men whose names were on the ledger was a political imperative that carried far-reaching strategic implications. And that refocused the coalition's targeting priorities into a strategic plan that was driven by vengeance.

13

Tradecraft

If you reveal your secrets to the wind, you should not blame the wind for revealing them to the trees.

—**Khalil Gibran**

THE GID WAS first and foremost a domestic counterintelligence and counterterrorist force. The mission of safeguarding the country, one surrounded by threats and instability, was challenging. Jordan was historically squeezed by outside conflict and the wars always managed, in one form or another, to seep into the kingdom. The country's 870 miles of frontier were desolate, porous terrain. The border areas were hellish under the desert sun where only the Bedouin and men with violent intentions dared cross.

Jordan had its share of homegrown extremists to contend with, as well.

To monitor subversion from within and from across its borders, the GID entrenched itself in most facets of daily life in the country, monitoring those who plotted to harm the country.[186]

There was a mystique of sheer cunning and unflinching ruthlessness surrounding the execution of the GID's internal security mission. That reputation was enhanced by the rumor mill and romanticized in fiction and in film, such as the movie *Body of Lies* starring Leonardo DiCaprio and Russell Crowe and based on the novel of the same name by *Washington Post* columnist David Ignatius. "Fear became its own currency," a former GID psychologist explained, "and the GID spent it wisely."[187] The GID always viewed counterespionage and counterterrorism as a cerebral exercise designed to seize the initiative and gain the advantage on the psychological battlefield.

The GID utilized foresight, force and manipulation to detect, deter and disrupt plots against the kingdom through displays of cleverness and manipulation that could have filled a dozen John le Carré novels. One famous operation targeted Palestinian terror mastermind Abu Nidal, and his organization in the 1990s. The GID knew that Abu Nidal was paranoid about traitors in his ranks and that he obsessed about which of his top lieutenants might be on the payroll of the CIA or the Mossad. The GID took advantage of this debilitating paranoia and established dummy bank accounts in the name of key operatives inside his organization. When details of these financial portfolios were strategically made available to Abu Nidal, his response was predictable and dozens of his top men were executed after kangaroo-court trials. Without firing a shot and without risking the life of an asset, the GID compromised the very fabric of one of the most dangerous terrorist groups in the world and rendered it ineffective.[188]

The Arab Spring and the Syrian Civil War stretched the GID's operational capabilities. The Western intelligence services estimated that there were thirty-six thousand foreign fighters in Syria and Iraq in early 2015. They came from eighty-six nations across the Middle East and around the world to fight for the Islamic State. The largest number of volunteers came from Tunisia; three thousand young men from the North Africa nation that gave birth to the Arab Spring trekked to Syria to participate in the holy quest for the Caliphate. Some, no longer able to venture to Afghanistan or Pakistan to fight the United States, traveled north toward Damascus to take up arms; some, many in fact, headed to Syria for a sexual romp and some ballistic excitement before heading back to their countries so that they could say they were there and boast of their combat prowess to family and friends. These thrill seekers were known as jihadi tourists.[189]

There were an estimated twenty-one hundred Jordanians who traveled north toward the Syrian frontier.[190] They were mostly middle class and educated; most had at least a bachelor's degree. They came from good families.[191] The GID knew many of them, and more important, knew their friends and associates, whom they corresponded with on chat apps, as well as on social media. Their mobile numbers—most had more than one phone—were easy to monitor, and the communications provided the GID with a telling look into what life was like on the ground—sometimes in real time—inside the war zone. The images and news the fighters exchanged with loved ones filled up gigabytes worth of memory in GID—and, ultimately, friendly intelligence services—databases. "If you wanted to know about any suspicious character in the region," a retired Middle Eastern intelligence officer commented, "you dialed +962, the country code for Jordan!"[192]

The homegrown jihadists were sloppy when it came to tradecraft. They communicated openly on their mobile phones and displayed even fewer filters when going on social media platforms. Names, locations, and small and seemingly insignificant comments were shared in phone calls, emails and text messages between the men in the trenches and their contacts back home and became an invaluable and seemingly endless source of intelligence for the GID; contact lists alone helped the GID assemble an elaborate and far-reaching quilt of information. The information became a gold mine for the coalition intelligence officers targeting high-priority targets inside Syria and Iraq. The material would be essential in the hunt for Moaz al-Kasasbeh's killers.

The GID had another invaluable reservoir of information on Syria inside the country. There were close to one million Syrian refugees in Jordan. Most were more than willing to assist the GID in its mission.

Jordan had a long history of receiving refugees. The Palestinians came in 1948 and then again in 1967. Kuwaitis—and Palestinians who lived and worked in Kuwait—came during and after the First Gulf War. The Iraqis who flooded into Amman in 2003 fleeing the Second Gulf War didn't come into the country carrying all that they owned on their backs like those before them but rather driving Mercedes and Bentleys across the borders, bringing cash and whatever else they could carry from Iraq. Yemenis arrived fleeing oppression, war and poverty; Eritreans, Darfurians and Sudanese came fleeing war and ethnic cleansing. Jordan didn't have the resources to accommodate all these refugees, and the influx of poor and desperate people pushed the country's weak economy to the edge of collapse. But the national policy was to never turn people away. The Arab Spring and the Syrian Civil War made an already precarious situation become a full-blown emergency.

The Jordanian border crossings with Syria were closed once the civil war began. Those fleeing the carnage had to find various spots along the separation fence where they could sneak through. Jordanian watchtowers positioned along the border were spread out at intervals of a couple of miles from one another. The frontier was dotted with military towers and isolated garrisons; each was issued with a numerical designation. One of the most popular places to cross was the Jordanian military tower located due south of the city of Daraa in southwest Syria, just five miles from the border fence on a hilltop in between Irbid and Mafraq in Jordan.

The tower south of Daraa was an isolated and sleepy outpost where hapless soldiers pulled guard duty. When the refugees came, Jordanian troops at the border were both ill prepared and ill-equipped to process the human swarm of desperate Syrians who streamed across the frontier bearing only what they could carry on their backs. Unlike their Iraqi counterparts years earlier who had applied for refugee status having driven to government centers in their air-conditioned luxury sedans, the Syrians walked across the desert and mountains in brutal heat, bone-chilling cold, monsoon-like rains and quicksand-like mud. Government forces were known to fire upon the long lines of men, women and children walking slowly on an attempted exodus toward freedom. "They used us for target practice," a thirty-year-old Syrian refugee who only gave his name as Salim explained from his corrugated tin hut of a house inside a refugee camp. "There were old men and women in our group. Assad's men are barbaric."[193]

Government snipers, the refugees knew, didn't want to simply kill those fleeing for the border—they wanted to maim disloyal subjects who sought a better life. The sharpshooters who preyed on the long line of escapees often aimed for the back, hoping to hit their targets in the spine. They went specifically after children. "If you want to kill a man, or you want to kill a child, you put a bullet in his head or his heart," a doctor treating the wounded along one of the Syrian frontiers commented. "They purposefully put the bullet in the lumbar [lower] spine so that the child would suffer. I don't have any other explanation."[194]

The International Red Cross and Red Crescent Movement (ICRC) and an army of NGOs set up shop along the border to care for the refugees. Hollywood celebrities and international soccer stars came, too, wanting to show their concern for the millions of displaced persons. The Syrians needed urgent medical care, food, water and shelter. Army tents were trucked in to set up ad hoc camps. The army provided emergency food and care—the intelligence case officers were the ones who provided the authorization to stay in Jordan. There was great concern that militants from al-Qaeda and other fundamentalist forces would smuggle operatives inside the throng of humanity pushing through the fence, so screening protocols were established along with GID and army intelligence officers deployed to the watchtower areas to interrogate the arrivals. Names and ID numbers were cross-referenced against databases and reference requests made to friendly intelligence agencies in the United States, Great Britain, Germany and France. GID liaisons to Iraqi and Saudi intelligence checked on Syrian men who might have crossed the border during the Gulf War to fight the Americans, or anyone known to have attended a madrasa—or training camp—in the tribal areas of Pakistan.

CIA officers from Amman Station, British specialists from MI6 also took the occasional two- to three-hour drive to the border crossing to interview special persons of interest.

There wasn't time to conduct lengthy interviews at the watchtowers—the area was flooded with humanity in dire need of permanent shelter, proper hygiene facilities and three meals a day. The weather was either skin-scorching hot or bone-snapping cold. The ICRC and the UNHCR* refugee assistance program did all they could to process the refugees and tend to the wounded and the sick. Buses were brought in to ferry those cleared for entry into Jordan as refugees to their temporary homes.

Whoever coined the phrase "if you build it they will come" must have had the Zaatari refugee camp in mind. Zaatari was named after the small farming village that had sat on an inhospitable patch of rock and earth for centuries and was located ten miles east of the city of Mafraq—an inhospitable plot of earth in the middle of Jordan's northern desert sixty miles north of Amman; Mafraq was a transportation hub connecting the king-

---

* United Nations High Commissioner for Refugees (formed in 1950).

The devastation caused by an al-Qaeda in Iraq suicide bomber in the grand ballroom at the Radisson SAS Hotel in Amman in Jordan on November 9, 2005. Sixty civilians were killed in a three-pronged attack ordered by Abu Musab al-Zarqawi; hundreds more were injured.

Photo: Jordan News Agency (Petra)

Left: Screenshot of Abu Musab al-Zarqawi taken from an al-Qaeda in Iraq propaganda video.

Photo: AFN/US Defense Visual Information Distribution Service

Below: US troops capture dozens of Sunni insurgents near Buhriz, Iraq, in April 2007. Many of the detainees would be housed in either Abu Ghraib or Camp Bucca—the ultimate recruiting grounds for what would morph into the Islamic State.

Photo: 5th Mobile Public Affairs Detachment

Insurgent prisoners seen here behind the wire at the Camp Bucca theater internment facility, the largest detention center in Iraq, as the security detail watches over them on a catwalk in 2009—two years before many in the facility's "alumni" would launch the Islamic State. Photo: 50th Infantry Brigade Combat Team Public Affairs

US secretary of state John Kerry arrives in Amman in June 2013 to discuss the deteriorating situation in Syria. Photo: US State Department Photo/Public Domain

The Zaatari refugee camp, seen from the air from a helicopter ferrying US secretary of state John Kerry in July 2013. **Photo: US State Department Photo/Public Domain**

The chairman of the joint chiefs of staff, US Army general Martin E. Dempsey, receives a salute from a Jordan Armed Forces Honor Guard during a planning visit in August 2013. **Photo: Department of Defense photo by D. Myles Cullen/Released**

On August 31, 2013, President Obama meets in the White House Situation Room with his national security advisors to discuss his decision to postpone a military strike against the Syrian government.

Photo: Official White House photo by Pete Souza

King Abdullah inspects military forces prior to the fall of Mosul and the formation of the international coalition against the Islamic State. Photo: Royal Hashemite Court

An RJAF F-16 pilot prepares to scramble during joint combat exercises with the US Air Force over Jordan.
Photo: Department of Defense photo by D. Myles Cullen/Released

President Barack Obama participates in a briefing on the campaign against the terrorist group ISIL in Iraq and Syria, held at US Central Command at MacDill Air Force Base in Tampa, Florida, September 17, 2014.
Photo: Official White House photo by Pete Souza

A US Navy F-18E Super Hornet is refueled by a KC-135 Stratotanker over Iraq after conducting an airstrike, October 4, 2014. Photo: US Air Force photo by Staff Sgt. Shawn Nickel

Moaz al-Kasasbeh receives his RJAF pilot wings from Prince Ali bin Al Hussein in 2009. Assigned to the RJAF's No. 1 Fighter Squadron, Kasasbeh flew combat missions against Islamic State targets throughout fall 2014 until he was forced to eject over Raqqa on December 24, 2014. Photo: Jordan Armed Forces-Arab Army

In this December 24, 2014, photo released by the Islamic State, a gleeful Abu Bilal al-Tunisi, a local Islamic State commander in Raqqa, seizes Lieutenant al-Kasasbeh from a riverbed, after the Jordanian pilot was forced to eject over Raqqa.

A video grab from the Islamic State's barbaric video of First Lieutenant Moaz al-Kasasbeh's immolation that was tweeted to the world on February 3, 2015. It is believed that the filmed execution was carried out weeks earlier.

King Abdullah confers with President Barack Obama on February 3, 2015, moments before the Jordanian monarch left the United States for Amman to handle the response to the murder of First Lieutenant Moaz al-Kasasbeh. **Photo: Royal Hashemite Court**

February 4, 2015: A convoy of Jordanian security vehicles departs Swaqa prison near Amman following the execution of al-Qaeda in Iraq terrorists held by Jordan. **Photo: Muhammad al-Kisswany**

King Abdullah embraces Safi al-Kasasbeh, Moaz's father, in a condolence call to the family's village near Kerak on February 5, 2015. **Photo: Royal Hashemite Court**

A Jordanian woman holds a candle in memory of Moaz al-Kasasbeh during a vigil in Amman.

Photo: shutterstock.com

RJAF aerial footage of the concentrated air strikes mounted against Islamic State targets in Raqqa in response to the release of the video depicting Moaz al-Kasasbeh's execution.

Photo: Jordan Armed Forces via Jordan TV

Right: A Jordanian military poster reads, "The Home of Martyrs: Moaz Martyr 2475" (the number of Jordan's war dead).

Photo: Jordan Armed Forces-Arab Army

Below: King Abdullah meets with CENTCOM commander General Lloyd Austin (center) along with US ambassador Alice Wells in February 2015 as the effort to punish those responsible for Moaz al-Kasasbeh's murder becomes a top coalition priority.

Photo: Royal Hashemite Court

King Abdullah confers with GID director Faisal Shobaki (right) at GID headquarters to discuss the war against the Islamic State. **Photo: Royal Hashemite Court**

A US State Department Diplomatic Security Service Rewards for Justice wanted poster for Islamic State leader Abu Bakr al-Baghdadi, offering a $25 million reward for anyone who brings about his capture.

Photo: US State Department/Diplomatic Security Service

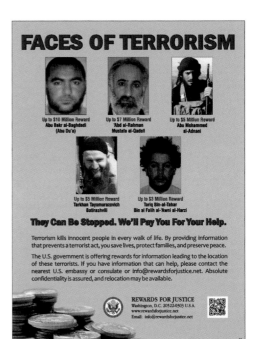

A US State Department Diplomatic Security Service Rewards for Justice wanted poster featuring Abu Omar al-Shishani (lower left) and Abu Mohammed al-Adnani (upper right).

Photo: US State Department/Diplomatic Security Service

People gather near the Le Petit Cambodge restaurant on Rue Alibert in tribute to victims of the November 13, 2015, attack in Paris at the Bataclan theater. **Photo: shutterstock.com**

A US Navy pilot assigned to Strike Fighter Squadron 131 taxies an F/A-18C Hornet on the flight deck of the aircraft carrier USS *Dwight D. Eisenhower* in the Persian Gulf in September 2016, around the time of the strike against al-Adnani's lair near Aleppo. **Photo: US Navy photo by Petty Officer 3rd Class Nathan T. Beard**

Gunners from the 101st Airborne Division conduct a night-fire mission in support of the Mosul offensive on December 6, 2016. **Photo: US Army photo by Spc. Christopher Brecht**

A Royal Australian Air Force F-18 Hornet flies a combat mission over Mosul during the offensive to liberate the city from the Islamic State. **Photo: Australian Ministry of Defense**

Iraqi security forces operate in the devastated streets of Mosul on June 22, 2017.

Photo: Corporal Rachel Diehm, 2nd Brigade Combat Team, 82nd Airborne Division Public Affairs

Right: Female trainees in the Syrian Democratic Forces (SDF), representing an equal amount of Arab and Kurdish volunteers, stand in formation at their graduation ceremony in northern Syria, August 9, 2017.

Photo: US Army photo by Sgt. Mitchell Ryan

Below: US Army Special Forces from the 5th Group assemble at the forward desert operating base in al-Tanf, Syria, in fall 2017.

Photo: Staff Sgt. Jacob Connor, 5th Special Forces Public Affairs Office

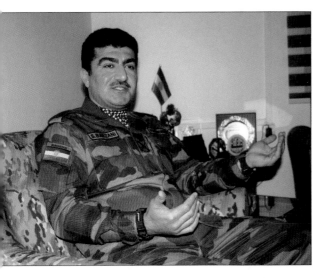

General Sirwan Barzani, Kurdish Regional Government commander of the Peshmerga in Sector Six, southeast of Mosul. His forces, as of fall 2019, are still involved in anti–Islamic State operations in the area.

Photo: Samuel M. Katz

US Army general Joseph L. Votel, commander, US Central Command, greets a local child during a tour of the city of Raqqa, January 22, 2018.

Photo: US Air Force photo by Tech Sgt. Dana Flamer

First Lieutenant Moaz al-Kasasbeh's flight jacket, peaked cap and photo are displayed at the Martyrs' Memorial in Amman. **Photo: Samuel M. Katz**

The author interviews King Abdullah in Amman, September 2018. **Photo: Royal Hashemite Court**

dom to Iraq and Syria. The patch of land ten miles east of the city of Mafraq at first handled a few hundred people housed in tattered military canvas tents. Soon tens of thousands of people would flow into the sandy square. Prefab huts with corrugated tin roofs were hastily assembled; water was trucked in and sewage lines were hastily dug. Babies began to be born inside the camp.

The camp defined paradox. Although intended as a temporary fix, it had the feeling and all the trappings of being a permanent landmark. To many of the close to one hundred thousand residents inside the concertina fencing, Zaatari became a hopeless dustbowl of agony. The UNHCR did its best to provide an array of services and benefits to help the refugees, but the challenges were daunting. Mohammed,* a UNHCR liaison officer and former battalion commander in the Jordanian military, was one of the few permanent staffers who were to make sure that the needs of the camp residents were met. Children had to be schooled, the sick needed care and people needed to make money. Some were permitted to work outside the camp; others were able to leave the barbed wire and find work and housing in Amman, or in one of Jordan's other cities.

The Syrians camouflaged the scars that were etched on their psyches with remarkably stoic resilience. The refugees refused to succumb regardless of the hopelessness. A mercantile boulevard ran down the center of the camp where everything from pots and pans to wedding gowns were sold; there were toy shops for the kids, and stalls that sold amusements for the adults. The shopping promenade was sarcastically called the Shams-élysées (*Shams* being the Arabic for Syria).

The smell of falafel or honey-glazed sweet treats frying was inescapable. But so, too, was the smell of overflowing sewage and human waste. Sometimes, because fate always dishes out new challenges to those in the greatest need, vicious sandstorms swept across Zaatari and turned the patchwork of huts into an

---

* Family name concealed for security considerations.

airless hell of dust and sand. The walls and barbed-wire concertina around the camp were, of course, a permanent reminder
to the refugees that their status was precarious and uncertain.
Jordanian police and gendarmerie forces patrolled the perimeter and had the final word on who got in and out. Zaatari was
far from home, but each man and woman held inside the fencing realized that they were still alive, and as Syria ripped itself
apart and thousands died each day, that was all that mattered.

Ultimately close to half a million refugees would pass through
Zaatari.[195] It remains one of the largest towns in the country.
Two other camps—Mrajeeb al-Fhood and Azraq—had to be
built to handle the overflow at Zaatari. Babies were born in the
camps, couples were wed and the elderly died. Confinement,
isolation and hopelessness could not stop life and death.

The refugees were grateful to be in Jordan, and they were eager
to do whatever they could to help Jordanian intelligence when
summoned for information. Some of the refugees cooperated as a
sign of gratitude. Others were rewarded with money or an offer
of employment. Some simply did it for a weekend in Amman or
Aqaba with their wives and a chance for an hour or two of the
kind of adult intimacy that made them feel human and that was
impossible to find inside a one-room shack with their children
and extended family.

There were also Syrians inside the camps who peddled information. Being an informant was a thriving profession in a
secret-police state, and once in Jordan, the old hands found new
customers. The names and faces of suspect men fetched a top
price in the intelligence market, and there were always those
willing to pay. The GID was wary of such transactions because
the crafty ones sold overhyped or inaccurate information. These
assets were expelled from the GID list of suppliers, outcasts in a
society where a reliable personal reputation took years to build,
and word of one bad tidbit spread at hyperspeed.

The mobile telephone was the lifeblood of the intelligence

business, and it was a psychological and physical lifeline for the refugees. The fighting in Syria demolished ancient wonders and turned cities into wastelands, but cellular towers managed to survive the constant combat. When the refugees arrived in Jordan, most brought whatever they could carry along with ID of some kind; property owners brought deeds with them. But everyone had a phone, either a Chinese-made smartphone or one of the many higher-end options from Apple or Samsung. The smartphone was the only way for the refugees to stay in touch with those still in harm's way. It was a way to maintain family ties, preserve photos and keepsakes, and share news of the day with loved ones torn apart by the fighting. The refugees feared that the SIM cards from the two Syrian carriers—Syriatel and MTN—were monitored, so they switched to one of the many Jordanian GSM mobile carriers—Zain, Orange Jordan and Umniah. It was good for the GID. Those calls were relatively easier to monitor and the SMS and phone logs easy to acquire.

There were very few refugees in Zaatari from Raqqa—the zip code that the GID was most interested in. Those fleeing the butchery in Raqqa or Aleppo made their way fifty miles north toward the Turkish frontier, or they tried their luck westward, toward a port in Syria and the hope of a rickety ship to take them to somewhere in Europe where they could find safety for their families; according to the UNHCR, close to a million Syrians made their way to the nations of the European Union.[196] The refugees in Jordan primarily came from Daraa, where the Syrian Civil War began, from as-Suwayda in the south and from Palmyra and Deir ez-Zor closer to the Iraqi frontier, but many had family and friends in Raqqa and Aleppo, and they knew people who could be useful to the coalition intelligence services.

The GID was also Jordan's primary foreign intelligence service working to protect the national interests. Most countries split the responsibility of internal security and overseas espionage between two agencies—the United States had the CIA and the FBI, Britain had MI6 and MI5, and Israel had the Mossad and the Shin Bet—but not the GID. Splitting the spy services into two separate but equal components was viewed as prudent to many who feared that one agency in control of too much information would become all-powerful. But when the GID was formed in the nascent days of the country, King Hussein wanted to keep the spies under one roof, under one centralized command: a singular service with an all-reaching mission inoculated from the interagency bickering and miscommunications that rival services had to contend with. The JAF's military intelligence arm was the GID's only true competition in the espionage business, and its focus was on the military capabilities of neighboring states and not the hybrid warfare threats from enemies near and far. Former CIA director George Tenet, when interviewed for the documentary *The Spymasters: CIA in the Crosshairs*, illuminated the operational dilemmas involved in the espionage game when, reflecting about 9/11, he said, "If you can't make the swivel between foreign and domestic, you're gonna get hurt."[197]

The GID spearheaded the intelligence-gathering effort against the Islamic State. Iraq was historically of major importance for Jordan. Iraq and Jordan shared a common history and often referred to each other as cousins. The Iraqi kings were Hashemites, after all. Iraq's King Faisal I and Jordan's King Abdullah I, who led their new mandates after the First World War, were brothers; King Faisal II, killed in a coup in 1958, and King Hussein of Jordan were cousins. And, historically, Jordan was an easy—and frequent—intelligence target for Saddam Hussein's intricate army of spies and saboteurs. In 2003 the GID arrested a network of Iraqi spies who had plotted to poison water supplies used by American forces stationed near Zarqa; the Iraqis also planned to firebomb luxury hotels in Amman where American officers and government officials stayed.[198]

The GID officers who analyzed developments in Baghdad and who ran agents in the provinces were true subject-matter experts in all things Iraq. The Americans and allied services relied on the GID's expertise and cultural intimacy with the men that now swore their allegiance to al-Baghdadi and the Caliphate.

The tribal connection was key to the GID's ability to penetrate al-Baghdadi's terrorist underground. The tribes in Jordan all had connections that transcended national boundaries and blended into Iraq. The tribes that shared connections across frontiers were known for the martial skills, their code of honor and their loyalty. "The tribal bonds are more important than religion and nationality," a military intelligence officer named Nasser* explained, "and those bonds certainly override any affiliation to any terrorist organization, even one that promises an Islamic State stretching to the four corners of the earth."[199] The tribe protected one another and, should a GID officer need shelter behind enemy lines or access to someone connected to the Islamic State inner circle, he knew that his blood brothers would never betray him.

The GID's Baghdad Station operated with the tacit support of the Iraqi government.[200] GID officers cultivated assets, established networks of informers and recruited reliable sources to be the service's eyes and ears. For the first time in the service's history, the GID went on the offensive.

Alexis Debat, a former French defense ministry official, was quoted saying that "any network the GID puts its mind to destroying is gone."[201] It was an overly simplistic compliment made in the backslapping euphoria that followed the joint Jordanian-American operation that killed Abu Musab al-Zarqawi. The pursuit of Zarqawi took close to three years and about sixty Jordanians and countless Americans and Iraqis were killed in the many months that Zarqawi remained free. Good old-fashioned human intelligence sealed al-Zarqawi's fate. Jordanian officers used informants, dead drops, and the cloak and dagger that was

---

* A pseudonym.

essential in running an agent to pinpoint the precise location
of the safe house five miles north of Baghdad where Zarqawi
was hiding out.

The network that Zarqawi built was constructed on a Kevlar
foundation of fanatics and powered by an unquenchable thirst for
violence, but it lacked the organization to shield the principals
from outside penetration. In the days following the creation of a
council of clerics that would serve as the seed for al-Baghdadi's
future Islamic State, the first order of business was to create a
counterintelligence force that would hunt down and extermi-
nate all enemies foreign and domestic.[202]

Samir Abd Muhammad al-Khlifawi was the name of the man
who built the Emni, the Islamic State's counterespionage and
counterintelligence apparatus, from a few scribbles on a note-
pad into one of the Middle East's most effective and brutal secret police forces. Al-Khlifawi was tall and soft-spoken with a
gaunt appearance and a mild manner about him; a white beard
concealed a boney jawline. He had been a colonel, a counterintelligence military officer in Saddam Hussein's Directorate
of General Military Intelligence, who was schooled by the best
spy hunters in the former Warsaw Pact and the most vicious
minds in Iraq's Baath Party to ferret out traitors, saboteurs and
scapegoats, and to make examples of them and their families.
An Iraqi patriot, he had been indoctrinated into the belief that
extraordinary and inhumane measures were necessary to preserve the state and its leader.

Samir Abd Muhammad al-Khlifawi lost his state and his
leader following the American-led invasion. Paul Bremer's foolish decision to disband the Iraqi military and civil service denied
al-Khlifawi a pension and a future. Bitter and vowing revenge,
al-Khlifawi managed to meet Abu Musab al-Zarqawi and became a key officer in al-Qaeda in Iraq. Once underground,
al-Khlifawi was known by the nom de guerre of Haji Bakr.[203]

Even utterance of his name was forbidden, though. As an intelligence officer for Zarqawi, he held one of the most secretive posts in the terror group.

Haji Bakr was ultimately captured by Iraqi and American forces and was a prisoner at the Camp Bucca and Abu Ghraib incubators of terror, teaming up with the likes of al-Baghdadi and Abu Abdulrahman al-Bilawi. Haji Bakr escaped prison along with the men who would form the foundations of the Islamic State in 2010 and began to blueprint the design of what internal and external security inside the Caliphate would be like. The civil war in Syria was an opportunity to fast-track the ambitious plans of turning a religious idea into facts on the ground.

Haji Bakr considered himself to be a masterful spy chief and modeled himself on Markus Wolf, the legendary East German Stasi head who became the inspiration for Karla, John le Carré's KGB mastermind in the Smiley trilogy. In fact, Bakr designed his police state's infrastructure as a carbon copy of the East German Stasi. The Stasi, after all, was a close ally of Saddam Hussein's intelligence service and instrumental in helping the Baath leadership transform Iraq into a police state.[204] Haji Bakr called this force the Emni, from the Arabic word *emniyyah*, which meant trust, security or safety.[205]

The Emni was an all-powerful force inside the Caliphate responsible for everything involving the organization's security. It was the Islamic State's military intelligence arm, providing fluid battlefield intelligence gathering to commanders in the field, and advanced espionage operations inside Middle Eastern countries where the Islamic State was strong and intended to conquer. The Emni also maintained active espionage operations to undermine and deter coalition capabilities, as well as an active effort against the Syrian regime. But because nothing in Middle Eastern wars was ever defined by logic, Emni agents also maintained covert ties with President Assad's spies, the Iranian MOIS and Hezbollah's security arm.[206] There was also an Emni liaison desk that maintained contact with rival Islamic fundamentalist groups around the world. "The bad guys always found a way to talk with one another," a GID officer said. "They were pragmatist above all else."[207]

The Emni ran networks of agents inside Turkey to make sure that the pipeline of arms, equipment, cash and fighters flowed continuously into the Islamic State. They also sent operatives posing as refugees into the camps in Turkey and Jordan to make sure that vital information was not being leaked to enemy intelligence services. The Emni recruited and deployed foreign fighters in the Islamic State ranks for intelligence-gathering and special operations missions in their home countries.[208] The Emni also oversaw all thefts of oil, food and antiquities to make sure that the foot soldiers of the fight weren't tempted to take anything for themselves; anyone who stole from the Islamic State coffers was summarily executed. The Emni weaponized all the Caliphate's propaganda—to use it as a weapon of fear and to prevent anyone from even thinking of betraying the cause.[209] And, of course, the Emni was responsible for all aspects of the sex trade inside the Caliphate. The clerics and the religious wise men on the ruling councils might have found Koranic verses to justify the systematic rape of girls and women of all ages,[210] but the administration of the mass rapes, abductions and slavery fell to the Emni. "And, like any criminal organization," a GID officer named Omar explained, "those at the top satisfied their own perversions first."[211]

Most of all, the Emni was a dreaded, fanatic and barbarically minded internal security and counterespionage force. Each foreign volunteer traveling to the ever-stretching frontiers of the Islamic State was vetted by Emni investigators. Men who flew on the European low-cost carriers into Istanbul's Sabiha Gokcen city airport and then made their way into Syria to fight, find a bride and earn a mercenary's wage were questioned for days by Emni specialists; the Q&A was done in Arabic, even though many of the men coming from France, Belgium and elsewhere could muster only a few words of the language that they had picked up at home or in prison. Referrals—a good word from a local Islamic State recruiter, or from a local imam, or even a copy of their arrest records—were examined with great care to make sure the volunteer was either a jihadi tourist or a religious mercenary and not an enemy plant. All volunteers were quarantined while questioned and checked out. The interrogations were thorough and harsh and often involved beatings.

Deep-cover Emni agents overseas carried out secondary background checks in the home countries. Haji Bakr was wary of the Belgian General Intelligence and Security Service or France's Direction générale de la Sécurité extérieure attempting to insert a deep-cover plant among those flocking to fight for al-Baghdadi.

Emni detachments followed the fighting units into battle, and into each town and village that the Islamic Station captured. Emni specialists pored over captured documents and records from the local Syrian secret police stations to assemble a loyalty profile of everyone who lived under their control. And just like every fanatic organization before them—from the Khmer Rouge to the Iranian Islamic Revolutionary Guard—the Emni targeted rich and powerful families, seized their homes and assets, and administered harsh treatment to the family heads as an example to everyone else. Thousands were executed without due process, trial or reason. Fear was everything.

Fear deterred anyone from thinking about trying to assassinate the Islamic State's leadership. Emni agents served as a Praetorian guard to al-Baghdadi and all the top religious leaders and military commanders who sat on the ruling councils. The Emni also hunted spies. And enemy intelligence operatives—real or imagined—were everywhere.

Haji Bakr was killed in combat, caught in the middle of a firefight near the Syrian town of Tall Rifat, some thirteen miles due north of Aleppo in January 2014.[212] His position was so secretive that the Islamic State propaganda machine made no mention of his death. Bakr was replaced by Abu Mohammed al-Adnani.

Haji Bakr had been an old-school spy chief. The organization was defined with a specific hierarchy. Emni operations were carried out locally, along the battle lines, and regionally, deep inside the Middle Eastern nations that were at war with the Islamic State. Structure ensured success. In this capacity, the Emni functioned like East Germany's Stasi or one of Saddam Hussein's many spy services—a cumbersome yet functional bureaucracy that successfully safeguarded the state. Al-Adnani had no use for structure or for the Iraqi intelligence service veterans that Bakr recruited to lead the Emni's various departments. He sacked many of the men from Camp Bucca and Abu Ghraib and replaced them with Tunisians and Moroccans, Chechens and Turks—men banished from their native lands who could never return home unless it was as a member of the Islamic State. The North Africans and Chechens were desperate. Al-Adnani was brilliant yet psychopathic. Together they created a lethal combination.

Al-Adnani rightfully saw enemies everywhere. He knew that he was in the crosshairs of the CIA, MI6, Assad's air force intelligence, the Russian GRU and, of course, the Arab services led by the GID. He was obsessed by the dreaded *aljasus*, or spies, operating inside the Caliphate, and he took measures to hunt them down. Emni agents were always stationed alongside the *hisbah*, or police, at impromptu checkpoints to examine fighters and civilians passing through territory the Islamic State controlled. Anyone found carrying large amounts of cash was suspected of being an enemy agent and taken to be interrogated. Anyone caught with a GPS device was suspected of being a spy. Mobile phones were confiscated and examined. Anyone found to have a brand-new model smartphone was immediately taken in for questioning, torture and likely execution. If a smartphone was found to be empty, or recently reverted to factory conditions, the suspicion was heightened. The Emni even employed children to eavesdrop on conversations in the street[213] and to look at which family bought a new car or spoke about plans to travel to Jordan or to Kurdish lines as a possible sign that they had sold information for cash. Emni agents had even infiltrated both the Syrian and Iraqi Kurds. "Most of the Da'esh security officers were veterans of the game," a GID officer explained. "They were either in the security services of Iraq or somewhere in North Africa or they were in a prison learning the trade one interrogation at a time. They were a formidable foe."

Emni personnel had remarkable autonomy in the territory they controlled. They were judge, jury and executioner of anyone they suspected of treachery. Informers were everywhere: in the market, in the mosque and even in the ranks of the men at arms—in their barracks, their mess halls, and even the homes they established for themselves with child brides and other women seized at gunpoint. Taxi drivers and merchants were recruited or forced to cooperate; virtually anyone who had contact with the outside world was always cast under a menacing veil of suspicion. In such instances it was quite common for people to fabricate information on the innocent or those with whom they had scores to settle, anything to avoid a bullet in the back of the head or a beheading. "Everyone was in a constant state of fear about being spied on," a former militant admitted. "It was impossible to discuss anything, even with members of the same ethnic group, out of worry that the conversation was being recorded."[214]

Anyone that the Emni suspected of working for a foreign intelligence service paid the ultimate price—guilt mattered little in matters of Islamic State justice. The executions always followed lengthy interrogations and torture, and were always held in public. The condemned were brought to the main squares. Ordinary citizens were forced to watch as the unfortunates were

bent forward and pushed to their knees before an Emni executioner disposed of them. Being shot in the head was preferred
to the other more gruesome forms of murder the henchmen
sometimes dished out to the untrustworthy. Terrified citizens
called Raqqa the "butcher's block."

The Emni's lust for blood was matched only by their unbri-
dled greed. The Emni controlled much of the cash that flowed
into the Islamic State from black market oil sales, smuggling,
extortion and sex trafficking. The Emni controlled the weekly
stipends of cash handed out to the men in the trenches; those
who made the pilgrimage to Syria and Iraq might have dreamed
of the glory of jihad, but they fought for a paycheck.

GID psychologists assessed that Abu Mohammed al-Adnani
was a psychopath. Counterterrorism Division analysts who wrote
reports that were shared with the Friends in Amman Station
and with the British hypothesized that the beatings he endured
while in prison turned him into a mental defect who was unable
to control his impulses; if the clinical evaluation was correct,
al-Adnani wouldn't have been the first man—or the last—to
lose his moral compass after being tortured by the secret police. The same reports diagnosed al-Adnani as being a paranoid
schizophrenic. Al-Adnani thought that enemy spies were everywhere. So, too, were Syrian agents, and Iranian operatives.
Al-Adnani was a byproduct of the political penal system that
produced victims who suffered from countless psychological ail-
ments and mental scar tissue. However, al-Adnani was correct
inasmuch the spies were all around.

The French took control of Syria following the First World
War. French—not English—was the country's second language,
and the nearly two million souls who lived in the confines of
greater Damascus, the oldest inhabited city in the world, identified more with the chic sparkle of Paris than with the romantic
Bedouin attachment to the desert. Syria was also cut across re-

ligious and clannish lines that were foreign to the tribal culture that existed inside Jordan and Iraq. The GID couldn't summon common contacts among the tribesmen and clans in Syria as it did with its connections in Iraq in Syria—those natural and blood links simply didn't exist.

Syria was a beguiling and frustrating espionage target for the coalition intelligence services. Somehow the Syrian population managed to survive forty-plus years of life inside Syria before tempers and hatreds erupted into full-blown civil war. Seven different intelligence agencies spied on virtually every aspect of Syrian life to locate and make examples of anyone brave—or foolish—enough to voice dissatisfaction with the regime. Government spies were everywhere. They listened in on phone conversations, intercepted the mail and even planted microphones inside the bedrooms of married couples; informants were recruited from all walks of life to point a finger at their neighbors, coworkers and even spouses. Paranoia was beaten deep into the collective Syrian soul; a late-night knock on the door was feared more than a cancer diagnosis. People were often taken away in the middle of the night only to be never heard from again. Keeping quiet and trusting no one became part of the Syrian molecular structure; the Western spy agencies—even the GID—found these traits, beaten into the national psyche over generations, difficult to overcome.

But Syria was *the* target. Mosul and other stretches of territory that abutted the Kurdistan Regional Government lines heading south along the Tigris River in Iraq might have been the Islamic State's banking and territorial epicenter, but Syria was the Caliphate's soul. The Islamic State command structure was based in Syria, and the men who led the lightning-fast military assault across the Levant felt safe in Raqqa and in the Aleppo vicinity. The shadow battlefield of Syria was unlike anything that the GID had ever encountered before.

On paper, Syrian terrain controlled by the Islamic State was impenetrable to spies—the Emni was too entrenched in the miserable reality of day-to-day life under the black banner control. But misery spawned animus, and those who suffered at the hands of the Islamic State needed no cajoling to assist those hunting for information. There was no shortage of men whose wives had been raped, women whose husbands had been executed on a whim in the name of the Prophet, and too many children killed in the indiscriminate fighting. Revenge was *the* most powerful—and reliable—reason for someone living in Raqqa to be recruited as a spy.

Spying against the Islamic State was all about the up close and personal: the eyes on a target and the one-on-one rapport between asset and case officer. Being a good case officer required being tough yet considerate and even kind, asking the right questions and not stopping until the right answers are given. A good case officer never wondered if he'd be able to extract information from an asset but rather how.

A good case officer was paternal. He worried about a source's safety, health and family's welfare even though if the source was captured, there was nothing that the intelligence officer could do to save him or his family. "How many kids do you have that Allah will look after?" was the kind of question a GID officer would ask of a man he recruited. A case officer who spoke the same language as his source and who prayed to the same God knew how to gain the trust of a source with cultural keys that provided a sense of brotherhood and safety even inside confines as dangerous as Syria under Islamic State control. The cultural keys also provided psychological insight into running an agent. Fear, weakness and hardship were to be manipulated with the proper religious connections and family assurances. Pressure was always easy to exert on people who were desperately afraid for their lives. But knowing how to talk to a man who was pious and who believed that Islam was being harmed by the khawarij, required finesse and cultural and religious intimacy. An agent had to lie and deceive to survive, but when reporting to his case officer, the agent had to be straightforward and honest for the information he supplied to be believed by those higher up in the chain of command. Spying was a dirty business.

The best information always came from a mole, someone planted deep inside the targeted group, but it was nearly impossible to tempt someone who swore allegiance to the Islamic State to betray al-Baghdadi. "They were untouchable with the sort of temptation an intelligence officer usually used on someone he wanted to recruit or turn," a GID officer named Marwan* explained. "We couldn't offer safety in another country to any of the foreign fighters because they had come to Syria from other countries in the first place. We couldn't offer the operatives money, because those psychopaths stole whatever they wanted to. They had weapons and could rifle through anyone's pocket anytime they wanted. We couldn't blackmail the fighters by uncovering acts of personal perversion, because they were proud of their sexual crimes. They abducted and raped girls as young as seven and women as old as seventy. Everyone has a vulnerability, a soft-spot, or even a self-defined resolve that can be exploited, but how does an intelligence officer tempt someone who has abandoned his soul? These men already considered themselves dead and waiting for their spot in the afterlife. They were incorruptible and virtually impossible to compromise."[215]

The CIA pledged its official support for the GID mission to eviscerate the Islamic State leadership, but they had other missions and interests in Iraq and Syria. The CIA had a reputation of overpaying for information—the expression "a trailer load of cash," a former State Department agent commented, originated from somewhere in the Langley dictionary. CIA case officers working with the Inherent Resolve task force, in northern Iraq, and elsewhere in the Middle East all had interests to try to re-

---

* A pseudonym.

cruit human assets inside the Islamic State. It would not have been a stretch for Syrians and others to be working for multiple intelligence agencies—sometimes from the same country—at the same time. The staging areas of Turkey, northern Jordan and the Kurdistan Regional Government were full of men and women speaking different languages, wearing 5.11 Tactical khaki trousers, eating power bars and looking for sources.

The CIA, the DIA and the NSA all had separate operations going on inside the Caliphate. Shipping containers full of cash were useful when searching for friends in the war zone and the Americans always came with money. When cash couldn't buy assistance or information, weapons were the preferred currency. But here, too, the covert confusion diluted the allied effort. In northern Syria, in Azaz, Marea and Aleppo, rebels supported by the CIA were fighting rebels who were armed by the DIA and the Pentagon; the bloody battles were fought between the Fursan al-Haq, or Knights of Justice, that Langley supported and the Kurdish-dominated Syrian Democratic Forces who were the benefactors of the soldier-spies at Joint Base Anacostia-Bolling at the DIA.[216] The war—let alone the intelligence-gathering effort—was virtually impossible to coordinate and control, especially since the battle lines changed daily, sometimes hourly. The map of the Caliphate shifted with each firefight.

Clandestine officers from the CIA did not venture out to war zones with a small entourage. The lessons from Beirut, Khost and Benghazi were clear: Agency officers traveled under heavy guard—specialists and shooters—when debriefing a source inside war zones. Security was everything, and the Agency and the other American intelligence services preferred technical means such as satellite imagery, drones and cyber warfare to the shoe-leather-slugging work of a case officer behind enemy lines.

Every espionage training course anywhere in the world always instructed future intelligence officers that the recruitment of human assets traditionally happens in conquered territory when individuals could be pressed and sometimes squeezed for information. HUMINT efforts were always easier in countries engulfed by war where the central government was weak or nonexistent.[217] Or at least they were on paper. The HUMINT end of the manhunt for Kasasbeh's killers fell on the GID.

Prior to each mission, senior GID commanders would meet with the intelligence officers before they departed on their assignments into harm's way. Operations were meticulously planned in order to not only achieve results, but to make sure that the men returned home safe and sound. Enormous care—and countless man hours of preparation—were invested to make sure that each assignment was planned in such a way that maximum security was assured. But even though the GID went to great lengths to train and prepare its officers for the dangers of the work, there were never any absolutes in the espionage business—especially when facing off against an enemy like the Islamic State and the trigger-happy Emni henchmen that were employed as counterespionage spy hunters. GID commanders were confident in the ability of their men, but supervisory officers worried when their officers were out on operational assignments; the bosses spent long sleepless nights on their sofas at headquarters nourished by a diet of nicotine and caffeine as they monitored communications with their forces in the field who faced unimaginable peril. The commanders hoped that they would remain invisible and silent throughout, recalling the Bedouin proverb that "only in complete silence will you hear the desert."

14

The Third World War

When the rich make war, it's the poor that die.

—**Russian proverb**

THE OTTOMAN TURKS were notorious for using brutal force as the language to convince the Arab populations under their control that their hearts and minds mattered little. Blind obedience was all that mattered to the Turkish pasha. The Europeans, primarily the British and French, were masters at delivering promises of political autonomy to the Arabs under their mandates in order to build alliances and foster quiet in the street; these promises, of course, were ones that London and Paris had no intention of ever honoring. The Turks and the Europeans produced a Middle East that was a patchwork quilt of shattered bones and broken promises.

The Americans were simply the most recent in a long line of superpowers attempting to rule the Arab, profit from the Arab and influence the Arab. The Americans did not break bones as a policy, nor did they draw artificial lines on a map. American promises came in cash—shipping containers full of freshly minted one-hundred-dollar bills.

The CIA had the reputation of throwing money around, but they weren't the only US government agency that tossed dollars into the air like confetti. The US Agency for International Development, or USAID, one of the world's largest benevolent organizations dispensing enormous amounts of money to help fledgling democracies and American allies overcome war, poverty and disease, to assist and expand existing government functions, had an annual budget of $27 billion. The rebuilding of Iraq was one of its largest and most ambitious projects; the money, believed to total close to $7 billion, was meant to stabilize towns and communities ravaged by the American-led invasion and the subsequent sectarian violence and pay to build up the police and other basic services for the Iraqi people.[218] There have been no sober estimates concerning how much it will cost to rebuild the areas of Iraq where ISIS carried out its scorched earth form of governance and warfare (it might be impossible to estimate the cost of rebuilding the cities of Syria that were devastated by civil war).

The spies and benevolent diplomats weren't the only ones trying to rebuild what had been destroyed with duffel bags full of new notes from the US Mint. Commanders of various elements of the Joint Special Operations Command possessed magical credit cards with seven-figure spending limits; majors and colonels could feed a village, buy an aircraft or even ship a warlord's daughter off to Disneyland without having to worry about the bean counters and discretionary spending. It was the American way.

One of the most ingenious—and effective—uses of American buying power came at the hands of the US State Department's Diplomatic Security Service. The DSS, as it's known, the State Department's security and law enforcement arm, protected the secretary of state, as well as non-head-of-state foreign dignitaries visiting the United States, such as the Dalai Lama and members of Great Britain's royal family. Small and flexible, the DSS was well suited for overseas criminal pursuits of terror suspects; special agents were assigned to protect US embassies and consulates around the world; these regional security officers (and assistant regional security officers) interacted with their local counterparts and often had access to more accurate and timelier intelligence than the CIA officers assigned to the various embassy stations.

In 1993 DSS created an online mechanism by which rewards could be given for information leading to the apprehension of wanted terrorists. Known at first as HEROES, the program ultimately became known as Rewards for Justice and advertised the names, faces, aliases and reward information for some of the most wanted men in the world; the information was published in a myriad of languages and printed on everything from faux dollar bills to matchbooks.

The program's greatest success was the February 1995 capture of World Trade Center bombing mastermind Ramzi Yousef in Islamabad, Pakistan; a member of Yousef's cell, working on a plot to destroy a dozen American airliners over the Pacific, saw Yousef's face and the offer for $1 million printed in Urdu on a matchbook and walked into the US Embassy in Islamabad. Other Rewards for Justice success stories included the arrest of Mir Aimal Kansi, the Pakistani national who shot and killed two CIA employees in the parking lot at Agency headquarters in Langley, Virginia, in 1993;[219] and information that led to the deaths of Uday and Qusay Hussein, sons of Saddam, in their hideout near Baghdad, in Iraq in 2003, after a tip revealed their safe house to JSOC and the 101st Airborne Division.[220]

The program receives millions of hits each year and countless tips from the four corners of the world.

On May 5, 2015, Secretary Kerry authorized the DSS and the Rewards for Justice program to offer bounties on the heads of four Islamic State leaders. Anyone inside the Caliphate who had

pertinent information leading to the apprehension or termination of Abu Mohammed al-Adnani and Abu Omar al-Shishani
could earn $5 million.* The information could be sent online,
via email or on a landline number registered in Tajikistan that
was forwarded straight to DSS headquarters in the DC area.

The State Department hoped that a $5 million payday, along
with the unadvertised benefit of possible relocation to the United
States, would be enough to lure someone with a tidbit to share.
It was a classic use of soft power to eradicate a hard target. The
US Department of the Treasury designated al-Shishani and al-
Adnani as "Special Designated Global Terrorists" in the fall of
2014. Earning such a distinction meant that the US government could freeze and seize a target's assets as well as the assets
of those who did business with him. "The Department of Treasury designation was one of the most powerful weapons imaginable to isolate a terrorist leader," a former deputy director of
a Middle Eastern intelligence service commented. "With the
stroke of a pen and a small press release, an Undersecretary of
the Treasury could cripple a complete terrorist enterprise not
by eviscerating the men in its ranks, but by rendering it pen-
niless and powerless. The Department of Treasury designation
didn't rip an organization apart like the shrapnel from cluster
munitions but made it rot from inside once weapons couldn't
be acquired, salaries weren't paid to the men in the field, and
millions of dollars skimmed off the top for a warlord's personal
take suddenly evaporating."[221]

The measures taken by the Department of the Treasury and
the Diplomatic Security Service were highly effective. Furor in
the terror financial markets was always a good thing, and it was
hoped that the economic warfare might stop the Islamic State's

---

* In the same May 5 communiqué, a $7 million reward was offered for Abd al-
Rahman Mustafa al-Qaduli, known by his nom de guerre of Abu Ali al-Anbari,
the deputy Islamic State commander and al-Baghdadi's deputy. A $3 million
bounty was placed on the head of Tariq bin al-Tahar bin al-Falih al-'Awni al-
Harzi, a senior commander from Tunisia. Abu Bakr al-Baghdadi was the most
valuable terrorist leader on the DSS want list, with a $10 million price tag.

funding, while at the same time offering an entrepreneurial soul a hefty reward for any information on the whereabouts of the Caliphate's most wanted. Yet these efforts were unilateral and done without the knowledge—let alone approval and support—of the hard power fighting the war on the ground.

The CIA and GID invested immense resources at great risk to develop leads that would connect a source or a phone number to one of the men they hunted. The offer of a sudden reward, and a full-fledged Madison Avenue advertising campaign to promote the bounty, forced the terrorist leadership deeper underground, not only eradicating months of work but warranting the creation of completely new networks. Amman and Baghdad Stations were not pleased with the development. Neither was the GID. The Islamic State leaders became even more paranoid. The men on the GID list changed their routines, more suspicious that those around them were traitors and profiteers.

The Emni was overwhelmed by the hunt for traitors. Nine days after the Diplomatic Security Service reward was offered online, a team of operators from the US Army 1st SFOD-Delta launched a daring raid to capture Fathi Ben Awn Ben Jildi Murad al-Tunisi, in al-Amn, eastern Syria.[222] Al-Tunisi was known by his nom de guerre of Abu Sayyaf, or "Bearer of the Sword," and was a Tunisian-born militant who had risen through the ranks to become Abu Bakr al-Baghdadi's most trusted deputy and the Islamic State's minister of finance; some reports likened him to Al Capone's accountant.[223] Abu Sayyaf was the Caliphate's money man and he knew where the funds were stashed, what the passwords were and which of the many illicit enterprises run by the terrorist army were the most profitable.

According to published accounts—not confirmed by the Inherent Resolve coalition—British operators from 22 SAS wore US fatigues and gear to tag along with their Delta comrades. With the help of Jordanian intelligence, 22 SAS monitored Abu Sayyaf's movements. Fearing traitors and coalition special operations intervention, the Islamic State financier traveled in a large motorcade made up of Emni special operations personnel—some who had experience in such work from their time in Saddam Hussein's Republican Guard. Abu Sayyaf's compound, near Deir ez-Zor, was a fortified villa surrounded by embankments and huts where the Emni protective contingent was billeted. The area was camouflaged from aerial intrusion, but it was impossible to shield the dozens of Toyota Land Cruisers and Hilux pick-ups from the prying eyes of reconnaissance satellites and drones.

JSOC and the intelligence services they worked for believed that in capturing Abu Sayyaf and the material in his possession, such as laptops and smartphones, they would grab the most sig-nificant treasure trove of physical intelligence seized from the grasps of the Islamic State since General al-Gharawi's comman-dos raided Abu Abdulrahman al-Bilawi's safehouse in Mosul eleven months earlier. MH-60 Black Hawks from the US Army 160th Special Operations Aviation Regiment (Airborne) flew the operators across the darkened desert to intercept Abu Sayyaf's motorcade in eastern Syria. The Islamic State leadership moved mainly at night.

Like most operations involving units such as Delta and the SAS, the roadside intercept was planned and executed with both precision and devastating firepower. But Abu Sayyaf's body-guards had no intention of surrendering their man and the ma-terial he possessed. A pitched battle ensued. Abu Sayyaf's Emni contingent fought doggedly, determined not to surrender an inch of ground. The firefight lit up the blackened Syrian night, though much of the battle was hand-to-hand and bare-knuckle. Abu Sayyaf was killed in the exchange.

The operators managed to pull rucksacks crammed with led-ger pads, personal scribbles and computers from Abu Sayyaf's smoldering Pathfinder. They also managed to secure Nasrin As'ad Ibrahim, Abu Sayyaf's wife, who was also known as Umm Sayyaf, in addition to an underage Yazidi woman, believed to be no older than fourteen, who had been held captive by the couple and used as a sex slave; the Sayyafs had seized numerous Yazidi women, murdered some of them and used the others for their own perverse purposes. Umm Sayyaf was an Iraqi national, and after questioning by intelligence officers from various coalition services, she was transferred to a prison run by the Kurdish Peshmerga in Erbil.[224]

Abu Sayyaf's death was a serious blow to Abu Bakr al-Baghdadi's Islamic State. With its financial operations paralyzed and the leadership wary of reward offers and commandos, the Caliphate's most wanted men dug in and went deeper underground. The war continued. Inherent Resolve air forces controlled the skies and the commandos owned the night. The spies worked the crevices of everything in between.

GID headquarters and Amman Station received inklings—nothing more than a rumor, of course—that al-Adnani was working on a special project. There were no specifics in the raw intelligence that was passed on from asset to case officer in a text message or a dead letter drop somewhere along a roadway leading to or from Raqqa; the word was that al-Adnani was working on a project to reshuffle the battlefield dynamic. But scuttlebutt wasn't actionable. The war continued and the regional and global powers all wanted their share of the spoils.

At just before dawn on the morning of July 23, 2015, a force of sixty heliborne commandos from Turkey's elite Özel Kuvvetler Komutanlığı (ÖKK), or Special Operations Command, landed in a valley near the Syrian village of Elbeyli-Ayyase, an Islamic State stronghold and staging area, approximately six miles from the border. The operators were the elite of the Turkish military—experts in long-range reconnaissance and direct-action raids—and they gained the high ground around the village in order to direct artillery and air strikes. Turkish armor crossed the border to support the operation, destroying buildings and possible Islamic State strongholds. According to reports—the Turkish media was forbidden on reporting on the foray—the operation netted one hundred Islamic State militants killed in action.[225]

That same day, Turkey allowed coalition combat aircraft to use air bases in the southern stretch of the country for tactical air strikes against Islamic State targets. Inherent Resolve commanders welcomed the decision, especially since those bases had been allowed only to launch refueling flights before. "Just in terms of physics, any time you can have your aircraft and your assets closer to your enemy, that's a good thing," a CENT-COM spokesman explained.[226] But Turkey's decision to go to war against the Islamic State took the coalition by surprise. Turkey was the gateway for virtually all of the foreign fighters who traveled from the four corners of the globe to fight for al-Baghdadi's Caliphate. More important, the emergence of the Islamic State weakened both Syria and Iraq—two of Ankara's primary rivals in the region.[227]

Twenty-four hours after the last Islamic State fighter was flushed out of Elbeyli-Ayyase, Turkish forces attacked positions of the Kurdish PKK, designated by the United States and the European Union as a terrorist group, in Iraq. Erdogan's entry into the fight terrified Bashar al-Assad. The Syrian leadership was not only concerned that Turkey was one of the only countries in the region which had the strategic and historical concerns—and the capabilities—to upend a military balance of power. From the Syrian perspective, the fluid and unpredictable battlefield was becoming dangerously crowded.

All eyes turned to Mother Russia.

Russian intelligence and special operations advisors arrived in Syria shortly after the civil war began in 2011. The emissaries were specialists, veteran communications officers and counterintelligence technicians. The Russians didn't advertise their overt support for their Syrian client, but they did little to hide the fact, either; pale-skinned Russians, faces burned red by the powerful Levantine sun, were hard to hide as they shuttled between the Syrian Defense Ministry and the Russian Embassy. Vladimir Putin believed in a policy of "rollback,"* turning back the clock to the day and age when there were just two superpowers with mutually assured capabilities, and Syria was key if Moscow was to ever regain its Cold War status.

This was, of course, no surprise to the American—and allied—intelligence communities. As far back as June 1, 1976, the CIA released an internal memorandum with the subject line: "Relations Between Syria and the USSR." The report began, "Syria maintains very close ties to the Soviet Union. These ties are based on Damascus' dependence on Moscow for the supply of military equipment and economic assistance, and on Syria's desire to be able, in a crisis, to turn to the Soviets for additional political and military support."[228] Syria's reliance on Moscow for political and economic support began in 1955, and in subsequent years, the two nations shared close ties through four Arab-Israeli wars; Soviet missile batteries protected Damascus during the 1973 Yom Kippur War, and Cuba—another client state of Moscow—sent commandos to fight Israeli forces in the effort to keep Israeli tanks from racing toward the Syrian capital.† Following the fall of the Iron Curtain and the end to the Soviet Union, Syria was Russia's sole remaining client state in the Arab world. The Russian navy enjoyed the benefits of a Mediterranean base at Tartus; Russian arms sales were brisk and plentiful. The Assad regime was Russia's only true foothold in the Middle East, and President Vladimir Putin was not about to let a strategic pawn disappear without a show of force.

---

* Other elements of Russia's "rollback," executed primarily by the GRU, later included social media bots and trolls to support ultra-right-wing nationalist extremists in Europe and the United States, and, of course, the successful efforts to assist Donald Trump in his march to the American presidency.

† North Korea sent MiG-21 fighter pilots to fly sorties against the Israeli Air Force in the 1973 war, and nuclear technicians from Pyongyang built a nuclear reactor at Deir ez-Zur that the Israeli Air Force destroyed in 2007.

Moscow's support grew as Syria's predicament on the ground worsened. Advisors were replaced by intelligence teams; Spetsnaz teams were inserted into the fight and Russian marines and naval infantry deployed to Syria's coastal areas. Russian air force planes resupplied Assad's forces with matériel and munitions. Russian pilots flew combat sorties in Syrian aircraft.

At first, Assad was reticent to beg Moscow for help, but by the summer of 2015, the Syrian President, in power since July 2000, had no choice. The Syrian army was entering its fourth year at war, and the military's capabilities were overwhelmed; senior commanders were dead, desertions were rampant and the air force was ineffective in providing close air support to its soldiers in the field, especially when the fighting was street to street and house to house. He asked the Iranians to coordinate Moscow's rescue. Tehran was only too happy to oblige. Major General Qassem Soleimani, commander of the Iranian Islamic Revolutionary Guard, received the mission to work out a deal with Putin. Soleimani, lionized by the Western media as the most dangerous spymaster in the world, commanded the Quds Force, the special operations and intelligence dirty-job arm of the Islamic Revolutionary Guard Corps. He was eager to use the Syrian situation to advance an Iranian-Russian alliance—such a partnership would ultimately save Assad's government and, at the same time, provide a superpower shield to the ambitions of the United States, the Sunni Arabs and even the Israelis, when it came to postwar Syria.

According to reports, shortly after Turkey entered the fray, Soleimani traveled to Moscow to meet with Putin and his military and intelligence commanders. Soleimani put the map of Syria on a large table that showed the most up-to-date problems facing Assad's forces. The Russians were alarmed by the graphics and the reality that Assad could lose his country. Soleimani was pure Machiavelli in his subtle manipulation. He assured the Russian leaders that there was still time to regain the initiative, while secretly using the Russian decision to act as a pretext to deploy thousands of Iranian and Hezbollah fighters into the warzone under the radar.[230]

Russian tanks and artillery batteries were unloaded off ships in Latakia in August 2015. The Russian air force took over much of Khmeimim Air Base near the coast, which soon became one of the busiest military airports in the world. Facilities were quickly built to accommodate one thousand airmen, and fortified bunkers were hastily dug to safeguard munitions. Runways were lengthened to accommodate Antonov An-124 and Ilyushin Il-76 transports. Russian fighter-bombers, including the Sukhoi Su-24s and 25s, as well as Su-34s, prepared for battle, as did Russian helicopter gunships.

In early September a Russian flotilla from the Black Sea fleet set sail toward the eastern Mediterranean. The die was cast.

On September 30, 2015, Russian aircraft began bombing Kurdish, Free Syrian Army, al-Nusra Front and Islamic State positions throughout the country. The Russians brought with them the same scorched-earth tactics that they employed in Afghanistan in 1979, and the Second Chechen War in 2000. Russian forces, primarily air and special operations units, delivered such massive and out-of-proportion firepower against rebel positions that, as one Moscow-based analyst commented, "The diehards died hard and the others changed coat."[231]

President Obama and his security staff, the men and women who did everything they could to remain out of Syria, received updates of the Russian buildup and realized that the United States had lost the geopolitical initiative in having a say in what postwar Syria would look like. "An attempt by Russia and Iran to prop up Assad and try to pacify the population is just going to get them stuck in a quagmire and it won't work," President Obama warned.[232] But words didn't matter—actions did.

Obama's projection that Russia was destined for military disaster in Syria was premature wishful thinking. Russia's intervention was done, as one report indicated, "on the cheap."[233] Putin was a cunning student of history and had, in his lifetime, witnessed two separate American quagmires in Afghanistan and Iraq. He was determined to commit the bare minimum with maximum results. Putin invested forty combat aircraft in the campaign, and only a few dozen transport and attack helicopters. Russia's Syria Task Force consisted of no more than six thousand troops—primarily special operations units and a few hundred hard-core mercenaries from a mysterious private military company called the Wagner Group; the Wagner Group has left its footprint and a body count in Africa and Ukraine and anywhere else the Kremlin requires deniable results.[234] The Wagner Group is believed to be owned by an oligarch who is close friends with Russia's Putin and has become a proxy force prosecuting Russian foreign policy when the use of actual Russian troops would cause an international furor; it is supposed to be able to field as many as three thousand to five thousand well-trained and highly motivated guns for hire who have made an impact in Ukraine, Central West Africa, Libya and, of course, Syria.[235] Putin didn't believe in repeating failed history, and he wanted to commit his forces in such a way that Russia would be indispensable in any talks to decide Syria's postwar future.

The GRU and their corresponding Spetsnaz hunter squads were on the offensive by the time the first Russian KAB-500R guided bombs fell on their targets. The Russians focused their special operations efforts on the Islamic State's military leadership, aiming to isolate key commanders and to terminate them. They pinpointed most of their operations—for tactical, as well as historical purposes—on al-Baghdadi's Chechens.

Russian aircraft pounded rebel-held lines day and night. Long-range bombers flew round-the-clock sorties from bases in Russia.

The battlefield was now dangerously congested and that risked a wider conflict that neither Washington nor Moscow wanted.

The chaos wasn't worth a Third World War. Once Russian forces became operational in Syria, the Pentagon and the Russian ministry of defense established a Memorandum of Operational Understanding to mitigate the possibility of both sides shooting at each other in midair: flight protocols were established to prevent collisions and engagements; specific frequencies were earmarked for missions; and agreements were made that aircraft would not fly at specific times and altitudes. In case of an incident, a twenty-four-hour red phone was established to prevent the escalation of the unthinkable.

The spies had no such memorandum.

In the field, without the benefit of uniforms and hotlines, the intelligence officers were on their own. The GID and the CIA were after many of the same men that the FSB and the KGB were after. The risk for bloodshed in such a lawless setting was high. The landscape was rife with treachery.

There was an immediate upside to the entry of Russian and large-scale Iranian and Hezbollah forces into the war. Islamic State commanders were no longer permitted to hunker down in their safe houses and bunkers in order to avoid coalition air strikes. There were battles to lead, artillery strikes to call in and tactical decisions to make.

On October 2, 2015, aircraft believed to have been launched from US aircraft carriers in the Mediterranean targeted a convoy of Islamic State vehicles racing from Raqqa toward Aleppo. The intelligence was that a local commander was heading to areas where Russian special forces were spearheading efforts to retake strategic crossroads vital to the movement of men and matériel. Four brand-new Pathfinders raced along a circuitous side route in a tight convoy protected by several pickup trucks fitted with heavy machine guns. The ordnance, powerful enough to turn a city block into a hole in the ground, turned the vehicles into a shattered contortion of twisted metal and flame. No one survived the bombing run, including the man being protected by the bodyguards: Abu Bilal al-Tunisi.

That night at Muwaffaq Salti Air Force Base, the pilots from No. 1 Fighter Squadron walked briskly from the dining hall to their operations room after dinner. The nights were getting colder and most of the pilots were just wearing their khaki-colored flight suits. The airmen turned on the television and boiled some water for tea and coffee, eager to see the news of the day and to make sure that their morning strikes over Syria hadn't made the headlines. Cigarette smoke filled the room.

Twenty minutes into the report on one of the Gulf station news services, there was a brief mention of an air strike in Syria near Raqqa. Among the men killed, the anchor explained, was Abu Bilal al-Tunisi, the man who pulled Moaz al-Kasasbeh out of the riverbed nearly a year earlier.

There were no cheers. No backslaps. It was a sober moment that brought back memories and highlighted the importance of the next day's mission they were preparing for.

The news network claimed that the Syrian Arab Air Force took al-Tunisi out. In fact, pro-Assad web sites released a posting of their own with the following:

To everybody all over the world, and to the people of Jordan in particular. This terrorist, the head of the Islamic Legitimate Body of ISIS, the terrorist who captured your Jordanian Pilot Muath Safi Yousef al-Kasasbeh, and the terrorist who your king promise he will avenge; the terrorist known as Abu Bilal al-Tunisi "Abu Bilal the Tunisian" This terrorist was killed in SyAAF [Syrian Arab Air Force] airstrike yesterday on Tishreen Garden by Jawad Anzour's high school in al-Raq'a, and his death was confirmed by SyAAF intelligence agents on the ground. Do not believe what your government-controlled media want you to believe; where were your governments from spotting ISIS convoys moving inside Syria and Iraq? Yet, for some miraculous reason, they can spot that our airstrikes only kill civilians?! There's so much that one can lie, and so much that one can believe without being called an idiot.[236]

There were no high fives in the conference room at the GID's Counterterrorism Division either. The officers spent much of the day trying to confirm the report that al-Tunisi had been killed; the news account only reaffirmed that he was dead and that the Syrian regime was claiming responsibility for the successful strike. The Russians also claimed al-Tunisi's head. News sources connected to the Russian government tweeted out mention of a Russian aerial attack.

It was politically smart and propaganda useful for the Syrians and the Russians to take credit for killing such a despicable figure, whose psychopathic grin personified the Islamic State fighters in the eyes of many around the world. It didn't matter that a coalition air strike took out the Tunisian commander; the GID didn't care who received the accolades for al-Tunisi's fiery end. The tradecraft that connected all the dots was never going to see the light of day anyway. All that mattered, one of the officers would later comment, "was that the bastard was dead and burned to a crisp."[237]

15

Say, Die in Your Rage

What lies ahead will be worse...for you haven't seen anything from us just yet.

—**Abu Mohammed al-Adnani**[238]

ON THE MORNING of January 7, 2015, Saïd and Chérif Kouachi, two French-born brothers of Algerian descent, walked into the Paris office of the highly controversial and provocative *Charlie Hebdo* satirical magazine and opened fire with automatic weapons.* They murdered eleven staff members before executing a police officer and then stealing a car that was used to flee the scene; they were cornered a day later and killed in a shootout with counterterrorist police. The Kouachi brothers were members of al-Qaeda in the Arabian Peninsula, or AQAP.

That same day, Amedy Coulibaly, a friend of the Kouachi brothers, laid siege to the Hypercacher kosher market also in the French capital. Coulibaly murdered four Jewish hostages before

---

* In 2011 the magazine published a Danish cartoon that depicted the Prophet Mohammed, and as a result, their headquarters was firebombed. The magazine used an unlisted address since that incident, but the two brothers found the publishing office's headquarters in the eleventh arrondissement.

being cut down in the police assault. Coulibaly was a member of the Islamic State; his wife, believed to be an accomplice, left France for Turkey a week before the attack, ostensibly to sneak into Syria and fight on behalf of the Caliphate.

France was a vital member of the international coalition against the Islamic State. As the former colonial master of Syria, its participation in the multinational effort was seen as vital in not only stemming the spread of al-Baghdadi's army across the Middle East but also playing a pivotal role in rebuilding a post–Islamic State and post-Assad Syria back up from the rubble. French aircraft were regularly seen over the skies of Iraq and Syria, and French special forces assisted Syrian rebels in the fight against al-Baghdadi's men. But French forces were fighting far beyond the Levant. They were in Libya. French Foreign Legion paratroopers were operating in Chad, and in Mali and in other African pieces of its former empire that were battling Islamic militants. But France faced dire security realities back home—internal challenges that had been ignored for too long.

Islam was the second-largest religion inside France, constituting just under nine percent of the country's population of some sixty-seven million people. The majority of the Muslims in France were from the former colonies of Algeria, Morocco and Tunisia. Those who emigrated from North Africa, as well as those who were first and second generation, were French in name only. They were never truly integrated into French society; they felt marginalized because of their surname, their complexion and their religion. They lived in the peripheries and projects alongside other immigrant groups and were ripe for radicalization and easy prey for cunning clerics and hypnotizing online sermons. The emergence of the Caliphate was attractive to many of these borderline individuals. According to the International Centre for Counter-Terrorism in The Hague, close to one thousand French citizens heeded the call to join the jihad in Syria.[239]

France's domestic intelligence service, the General Directorate for Internal Security, was overwhelmed by the number of leads that flowed into headquarters following *Charlie Hebdo* and Hyper-cacher attacks. The Direction générale de la sécurité intérieure, or DGSI, as it was known, was short on manpower, and it lacked the required number of experienced and reliable Arabic-speaking offi-cers to monitor known and suspected militants. "There were thou-sands of possible terror targets moving in and out of the country," a US State Department Diplomatic Security Service special agent based in neighboring Belgium commented. "It was beyond the grasp of most of the agencies in Europe to cover so many targets with 24/7 surveillance. It was simply impractical and unrealistic."[240]

It was the worst-kept secret in the world that the Islamic State was active in Western Europe—specifically in France and Bel-gium. The sermons preached in neighborhood mosques called for violence against the Crusaders and infidels; the graffiti on the walls of the slums and housing projects promised revenge and salvation. The message proliferated on social media sites such as Facebook and Twitter. The DGSI, the National Police, and the National Gendarmerie were well aware that there were Islamic State cells inside the country.

The Directorate-General for External Security, the French version of the CIA and MI6, had staff assigned to Syria and the coalition. The Direction générale de la sécurité extérieure, or DGSE, was considered one of the stronger Western intelligence services working inside the Arab world, especially in countries that France once ruled. DGSE officers interrogated terror suspects who came from France, and they were able to establish how fighters in Syria originated as petty thieves and drug pushers in the tenth arrondissement of Paris. But concrete evidence—real names, ac-curate addresses or plots on the radar—that could be turned into a proactive and preemptive surveillance operation was lacking.

The fall of 2015 was a critical period of combat for the Islamic State—both on the battlefield and inside the ruling council. The endless air strikes by the Inherent Resolve coalition fighter-bombers had taken a severe toll on Islamic State military capabilities and, most important, their economy. The bombings were relentless, and although a semiconventional counterguerrilla war could never be won solely by air power, the precision strikes were amazingly effective. The hodgepodge of anti–Islamic State militias and armies that received enormous American and coalition support were making advances on the battlefield, as well, shrinking the Caliphate's landmass; they were able to recruit new volunteers into the struggle to defeat al-Baghdadi's legions in every town and village they liberated. The Islamic State's ruling council found itself under tremendous pressure to hold the lines and to not relinquish another inch of territory. Russia's entry into the war changed everything. There was no way that the Islamic State army could hold off both the United States–led alliance and Russia's scorched-earth philosophy of warfare and do so on a fluctuating multifront landscape.

The increasingly precarious military situation brought to the surface a harsh religious, political and ideological rift that existed among the Islamic State's top decision makers. The battle lines were ethnic. The Iraqis and Syrians at the top viewed themselves as the true leaders of the organization. The North Africans, who were the most volatile and fanatical inside the group, wanted proper representation and a chance for future leadership consideration. The "others" consisted of the Chechens, Turkmen, Tajiks and fighters from the former Soviet Union's Asian republics. The infighting highlighted a power struggle that existed between Abu Omar al-Shishani and Abu Mohammed al-Adnani—two men who were vying for the all-powerful position of being chosen as al-Baghdadi's successor. Al-Shishani, the war minister, faced enormous internal criticism for not only not being able to withstand the multinational effort but for not going on the offensive. Al-Adnani thought that expanding the conflict to the "Crusader homeland," as Western Europe was referred to, would change the narrative and scope of the conflict and increase his standing in the ruling council's eyes.

Al-Adnani was the more ambitious of the two. In the end, ambition won. His men were already in place.

Abdelhamid Abaaoud was the twenty-eight-year-old Belgian-born son of Moroccan immigrants and commander of a highly active Islamic State terrorist cell. He grew up in Molenbeek-Saint-Jean, a section of Brussels that had become a North African ghetto, where it was as easy to find CD sermons from Raqqa as it was to buy a gram of hashish or a revolver or a kilo of prized Tunisian olives. The young men of Molenbeek, as the neighborhood was known, were either in jail, out of jail, or heading to jail, Patrick M.,* a former US State Department officer, claimed. "The police feared to patrol in the area," the special agent explained, "and it was easy for someone up to no good or escaping an Interpol warrant to find safety amid the locals."[241]

The local supermarkets and cafés that sold a taste of the old country also sold dreams of jihad and the promise of the Caliphate. Al-Adnani dispatched some of his department's most convincing salesmen to recruit impressionable young men into the ranks.[242] A stint in prison was considered a plus; Western intelligence officers called time behind bars a "jihadist accelerator." Men like Abaaoud who had criminal records and possessed explosive rage were considered ideal Islamic State material. Molenbeek produced close to five hundred such men. Syria was where they learned how to fight and kill.

According to reports, Abdelhamid Abaaoud spent a year in Syria in 2013 and swore his allegiance to the newly created Caliphate before returning to Western Europe. He recruited his younger teenage brother to join him in the fight. He was photographed in Syria loading corpses onto a truck.[243] He wore a characteristic *pakol*, the Pashtun hat favored by the mujahideen in Afghanistan, when he fought Assad's forces in Syria and the coalition-supported

---

* A pseudonym.

Kurds, Yazidis and soldiers from the Free Syrian Army and Syrian Democratic Forces.

The DGSI and Belgium's state security service, the VSSE (known as the Veiligheid van de staat in Flemish and Sûreté de l'état in French), had files on Abaaoud that were as thick as a Manhattan phone book. Abaaoud's name and mobile telephone number came up in the investigation of Mehdi Nemmouche, who shot and killed four people at the Jewish Museum in Brussels in May 2014. Abaaoud gave interviews to al-Adnani's *Dabiq* magazine taunting French and Belgian police. He didn't care that his face and name were known. In 2015 he was sentenced in absentia to twenty years in prison for terrorist offenses.

Abaaoud hid in plain sight, daring the authorities in Paris and Brussels to try to arrest him. With money funneled to him through Syria, he rented safe houses, acquired weapons and coordinated with al-Adnani's external operations unit in Syria to put the final touches on the Islamic State's boldest military operation on the European continent. Abaaoud and the eight men that were recruited for the operation communicated openly on Twitter, Facebook, Instagram, WhatsApp and Telegram.

As the months went on, though, French and Belgian authorities chased other leads in pursuit of other terror suspects that, the intelligence believed, were about to launch imminent strikes in Western Europe. Abaaoud's name lost its ticking-clock urgency. And then, on October 31, 2015, a bomb ripped through the baggage hold of Metrojet Flight 9268, a Russian chartered airline, shortly after it took off from Sharm el-Sheikh in the Sinai Peninsula; the flight was destined for St. Petersburg. The aircraft, an Airbus A321, had 224 people on board and disappeared from Egyptian radar somewhere over the desert. Everyone on board was killed. The Islamic State claimed responsibility for the bombing. It was the worst bombing of a civilian airliner since Pan Am 103 was blown out of the sky over Lockerbie, Scotland, in December 1988, killing 259 people in the air and

11 people on the ground. For al-Adnani's European operation, the Metrojet bombing was a brilliant diversion.

Napoleon Bonaparte was quoted as saying that "secrets traveled fast in Paris."[244] But it turned out that facts in plain sight moved at a snail's pace in the City of Light. Common sense warranted that French security forces maintain the highest state of alert in the national warning system, a color-coded alert forecasting tool that had been instituted close to forty years earlier by president Valéry Giscard d'Estaing and updated several times since. But on the night of Friday, November 13, 2015, fathers taking their sons to the Stade de France to watch a friendly soccer match between the German and French national teams were unaware that an attack was moments away from being launched. American exchange students about to eat at a trendy Cambodian restaurant were unaware of the threat level, as well. Concertgoers who waited months to see their favorite American band had no idea that Abu Mohammed al-Adnani had nurtured, supported and then green-lit a cell of home-grown veterans from the Syrian conflict to go on the offensive.

The attacks began at 9:20 p.m. Three suicide bombers blew themselves up outside the gates of the Stade de France in the suburb of Saint-Denis, where a soccer friendly match had just begun between the German and French national squads. One of the terrorists, a man identified as Bilal Hadfi, was French-born; the other two members of the terrorist group were believed to be Syrian nationals who snuck their way into Western Europe posing as refugees. A forged Syrian passport was found on the body of one of the bombers.

François Hollande, the French president, attended the match. He sat in a luxury box surrounded by security men who rushed him to safety the moment the first bomber exploded. The thunderous roar of the explosions ripped through the stadium as the near-capacity crowds were standing and yelling in support of their teams. Inside the stadium it was immediately evident that something was horribly wrong. Players rushed into their dressing rooms after scouring the stands for their wives and children. Pandemonium erupted as did a sense of foreboding.

Gunfire broke out in the tenth arrondissement five minutes later. Abdelhamid Abaaoud led the next three-man attack squad personally, as they opened fire and tossed grenades at diners and club-goers in the fashionable cafés and trendy eateries of the area. The three men armed with automatic weapons and wearing suicide vests emerged from a stolen SEAT sedan and opened fire outside the Le Carillon bar on 18 rue Alibert. They shot at anyone unfortunate enough to be in their way and tossed antipersonnel fragmentation grenades down the block and into the outdoor cafés, mowing down dozens of dazed and horrified diners. The gunmen then split into two groups in two separate vehicles; one, witnesses reported, had Belgian license plates.[245] The terrorists moved quickly up and down the street as they fired, and made it impossible for the responding police units to pin them down. Abaaoud's men returned to their SEAT and drove a few blocks toward Rue de Charonne and La Belle Equipe. There they shot their way inside and opened fire. They sprayed over a hundred rounds into the low-lit saloon. It was here that Abaaoud killed Nohemi Gonzalez, an exchange student from California, who was waiting for her friends to arrive for a celebratory dinner. Gonzalez was the only American fatality of the attack.

One of Abaaoud's men split from the group and blew himself up as police units closed in on his location. The detonation, it was hoped, would signal to police that the attacks in Paris had come to an end. Abaaoud and his partner then disappeared into the chaos and confusion, ditching their SEAT and heading to a safe house five kilometers away in Montreuil, an eastern suburb and hub of the capital's Arab community.

At 9:40, three men attacked the Bataclan theater on the Boulevard Voltaire in the eleventh arrondissement. The theater, one of the oldest in Paris, was hosting the Eagles of Death Metal, an American band with a strong European following. Some fifteen hundred fans were jumping up and down as the band played the song "Kiss the Devil." The three men, all French-born citizens of North African descent in their twenties, pulled up outside the venue and, dressed in black combat fatigues and armed with AKM 7.62-mm assault rifles, grenades and explosive vests, opened fire inside the crowded hall. Some witnesses heard one of the terrorists yell "God is Great" before opening fire; others thought the explosives and flashes were part of the pyrotechnic package that the band was known for.[246]

The Bataclan siege was the only part of the attack that night that became a hostage ordeal. Operators from the elite French National Police RAID unit (the French acronym for Recherche, Assistance, Intervention, Dissuasion; or Search, Assistance, Intervention and Deterrence) had already been mobilized because of the attack against the national stadium. They now surrounded the Bataclan along with police commandos from the Research and Intervention force, or BRI, in a controlled perimeter and were poised to storm the location to rescue whoever could be saved, while quickly terminating the terrorists.

Police teams stormed the location under heavy fire, and they rushed into a scene of unimaginable carnage. Blood and spilled beer were everywhere, creating a thick and slippery obstacle that the first responders had to move through. There were bodies on the floor, along with tossed-out handbags and smashed cell phones crushed under the weight of stampeding music fans who fought their way to safety. The hostage-rescue teams moved cautiously over the dead and wounded, even as they were subjected to intense rifle and grenade fire. A raging firefight ensued. One of the terrorists blew himself up after shooting it out with police and being cornered in a rear room. The other two terrorists detonated the explosive vests strapped to their chests as the police teams moved upstairs to secure theater offices and the balcony.

By the time the sun emerged the next dawn, 130 were dead and close to 500 had been seriously wounded. It was one of the deadliest attacks ever in Europe and the worst terrorist incident in French history.

Seven of the nine terrorists were killed that night. The two survivors, including Abaaoud, became the subjects of the largest manhunt in European history. Abaaoud and other members of his cell were cornered four days later in the Saint-Denis section of Paris, in a safe house that had become a fortified bunker and arsenal. When police commandos stormed the safe house, it soon turned into a seven-hour gun battle in which over five thousand rounds of ammunition were fired.[247] The bodies of eight militants, including Abaaoud's, were pulled from the bullet-scarred building, a foreboding indication that al-Adnani's European operation was larger than French—and Belgian—intelligence could have ever imagined, and was still far from being defeated.

The French declared a state of emergency immediately following the Paris terror attacks. Similar security precautions were initiated in Belgium. Although French air force bombers launched a ten-aircraft sortie against command and control targets in Raqqa, their largest of Opération Chammal (the name for the French participation in Inherent Resolve), France's real fight against the Islamic State wasn't in Syria—it was in Paris and in the slums of Brussels. Al-Adnani had been successful in both his tactical and political goals of expanding the battlefield beyond the Middle East and in trumping al-Shishani's status as the most capable and important field commander in the Islamic State.

Belgian security forces had been aggressive in their hunt for Islamic State terrorists in the capital—particularly those living in Molenbeek. Police raids became routine; arrests were made, and arms caches were seized. But on the morning of March 22, 2016, the remnants of the Abaaoud Islamic State cell struck Brussels. At just before 8:00 a.m., three men arrived together by taxi at Brussels International Airport. Two of the men detonated large suitcases crammed with explosives and nails at the check-in counter; a third bomber attempted to detonate his suitcase, but the device's mechanism was damaged by the after blast of the other two explosions. He disappeared in the chaos and confusion of the smoke-filled departure hall, walking over the bodies of the dead to reach an exit not cordoned off by police. Closed-circuit television cameras captured his escape,[248] but his face was concealed by a large and floppy black hat and he disappeared into the Muslim neighborhoods of the capital, where the Islamic State was wildly popular and where nearly all of the Belgian foreign volunteers to the war originated.*

Belgian authorities immediately shut down metro and rail service to the airport. The transit slowdown was meant to limit the means by which additional suicide bombers could reach their targets, but one was already on the metro. At 9:11, just as a train on the number five line departed Maelbeek Station, the younger brother of one of the airport bombers blew himself up in the middle car. The powerful yield of his explosive device ripped across the train and damaged a large portion of the station.

Thirty-two people were killed in the Brussels airport and metro bombings; 230 were critically wounded.[249] It was the worst terrorist attack in Belgian history.

The Paris and Brussels attacks caught the intelligence services fighting the Islamic State on the ground in Syria and Iraq completely by surprise. Al-Adnani's external operations apparatus inside the Caliphate had been successful in compartmentalizing the infrastructure and intentions he had assembled in Western Europe. Targeting al-Adnani was no longer just a Jordanian point of order and national revenge. It was now a French, Belgian and international priority of maximum imperative. The only question was how many more innocent people would have to die before al-Adnani—and the others—were confirmed killed.

---

* Mohamed Abrini, the "man in the hat," was arrested by Belgian security services on April 8, 2016, in the Anderlecht section of Brussels following an exhaustive manhunt.

BOOK FOUR

PAYBACK

16

Desert One

Let your plans be dark and impenetrable as night, and when you move, fall like a thunderbolt.

—**Sun Tzu**

IN 2003 AND 2004, in the aftermath of the American-led invasion of Iraq and the onset of the first insurgency, the Syrians facilitated an insidious conspiracy. Foreign volunteers from around the Middle East and beyond were permitted to travel through Syria toward the fight in Iraq. The men who came from North Africa and Arabia flew on direct flights into Damascus International Airport. They were processed at immigration by officials who helped them secure travel east; in some cases, a small envelope with cash was waiting for them. The men shrugged off their jet lag and, instead of visiting the Umayyad Mosque, one of the oldest and most revered in the world, the travelers hired taxis to take them to the barren stretch of desert north of the Jordan where the Syrian frontier and Iraq's al-Anbar Province converged.

No one traveled through the inhospitable desert unless the

Bedouin were involved and paid. The Bedouins were master smugglers, and they owned the desert; they knew every nook and cranny in the blistering moonscape, and they determined what moved in and out of their God-given land. According to Raja,* a GID officer in the CTD, "Smugglers made $30 bringing a single man into Iraq. On good days a smuggler could make as much as $500. The traffic was that intense and the business was a staple of the Bedouins' income."[250]

Through a source, unspecified and classified to this day, the GID tapped into this trafficking network. The GID received almost daily updates with specific and detailed information on the vast majority of the foreign recruits—names, passport numbers and passport photos—and the files were shared with the CIA and the other Western services with fighting men in Iraq. But the United States opted not to interdict the pipeline of manpower and armaments into Iraq because it didn't want to escalate the conflict into Syria and become embroiled in yet another Middle Eastern ground war.

History repeated itself a decade later when the Islamic State used the exact same underground railroad in the reverse direction, traveling from Anbar west into Syria. All traffic ventured through a dusty crossroads smack in the middle of nowhere called al-Tanf.

Al-Tanf was both Mars and Devil's Island rolled into one. A parched dry slab of flat desert punctuated by a junction and a small police garrison that Assad's customs service used to collect tribute on the people, machinery, weapons and narcotics that crossed in between Iraq and Syria. "The officers deserved to be corrupt and to steal their share," an American Special Forces operator commented, "because just serving there was like falling off the face of the earth into a purgatory of dust and fire."[251] The summer was a pure and unrelenting inferno. It was as close to hell on earth as could be found on this planet, prompting the

---

* A pseudonym.

locals to joke that not even the scorpions were foolish enough
to call this desert home. Temperatures in the shade routinely
topped 120 degrees Fahrenheit, though there was little shade
to be found. There were no trees, no cover—just wide-open
desert—little to shield someone foolish enough to be out in the
middle of nowhere from the face-scorching force of the powerful winds. Blinding sandstorms were common; the odd rain-
storm that passed through turned the sand into a grainy, sticky
projectile that whipped faces and clogged engines.

Al-Tanf was ravaged by weather and downright neglect. The
customs station consisted of a small fort and a series of single-
story structures inside an encampment sloppily painted with
coats of whitewash. A steel arch structure hovered above the
main gate. Before the civil war, a large canvas mural depicting
Bashar al-Assad and his father, Hafez, the former Syrian strongman, hung from the archway before the entrance. Concrete pill-
boxes that once controlled access in and out of the camp were
abandoned. The buildings probably dated back to the French
Mandate—and possibly even the Ottoman Turks. The customs
officer who commanded the facility was always a member of
a family with government connections or a Baath Party insider; the assignment, a cash cow for an officer to accept kickbacks and bribes, was considered a reward for blindly supporting
the regime. Only agents who would fight to the death to preserve the Assad family reign were stationed at al-Tanf. It sat on
the main—and only—highway connecting southern Iraq and
Baghdad with Syria and Damascus, and was considered one of
the most strategic outposts for maintaining the Syrian regime's
contact with Baghdad and Tehran. The garrison in al-Tanf held
their positions through continued Islamic State assaults for over
a year. In most cases, as the Islamic State marched west, the
men in black simply bypassed the fortified position. But in May
2015 it fell to the Caliphate. The black flags and banners were
hoisted above the main entrance and murals depicting Bashar

al-Assad in uniform were painted over and then replaced with scriptures from the Koran.

CJTF-OIR coalition commanders viewed al-Tanf as a zeroed-in-yet-prime block of invaluable real estate. Operationally it was a kill zone: fighters flooding in from Iraq, as well as those fleeing the combat in Syria, could easily be targeted as they drove along the desert highway. Allied aircraft flew hundreds of sorties targeting Islamic State vehicles that moved through al-Tanf between Iraq and Syria. Al-Tanf was only eleven miles north of Tower 22, the most remote and desolate of all Jordan's border markers with Syria. Tower 22 was an invaluable intelligence-gathering post to monitor Islamic State cellular communications. The Rukban refugee camp and its overflowing and always growing population of forty thousand–plus refugees was situated nearby; there were always caravans of cars, some barely functional, trying to reach the camp and relative safety sharing the same Damascus–Baghdad highway with the terrorists. Unlike the Zaatari camp in northern Jordan that boasted refugees primarily from Daraa, Damascus and Deir ez-Zor, Tower 22's location covered a wide swath of traffic from both Syria and Iraq, and that made the location an intelligence gold mine. People who were in Raqqa and Aleppo only days earlier became ideal sources of intelligence.

It took less than an hour to reach Rukban from Tower 22. The roads were monitored by Islamic State units, and ambushes and IEDs were frequent obstacles, so the intelligence gatherers traveled on desert paths that only the Bedouin could negotiate—often at night. Jordanian guides and translators accompanied every allied mission. When sources were interviewed, the nuance of language and cultural understanding was of critical importance.

Jordanian and American special operations units routinely launched reconnaissance and, later, direct-action raids against Islamic State targets moving on the desert highway. The operations were covert and high risk. Islamic State unit commanders in the area offered their men a hefty bounty for the capture of an American or Jordanian spy.

In the early months of 2016, seizing upon the intensified aerial effort against the Islamic State in the wake of the Paris and Brussels terror attacks, the combined Russian, Turkish and coalition aerial assault against Caliphate strongholds intensified. The Iraqi military, complete with its American advisors, slowly regrouped and began seizing ground that the Islamic State had held for two years. President Obama boasted that the tide had turned, and that "their [the Islamic State] ranks of fighters are estimated to be at their lowest levels in about two years, and more and more of them are realizing that their cause is lost."[252] As summer arrived, the United States increased the number of combat personnel assigned to Iraq, ostensibly to assist the Iraqi Army to prepare and execute an offensive that would retake Mosul and deliver a crippling blow to the Islamic State.[253]

At the very end of March 2016, as details of al-Adnani's sleeper cells in Western Europe were being discovered by investigators tiptoeing through the crime scenes in Brussels, Syrian rebel forces supported by allied aircraft seized al-Tanf. It was a turning point in the war. A steady stream of helicopters began to land on the flat desert sand within hours of the fort's capture.

In days, al-Tanf became a fully operational coalition base deep inside Islamic State territory. The Americans rushed engineering and construction supplies there. The convoy of supply and transport vehicles traveled from Tower 22 under the air cover of coalition fighter-bombers and helicopter gunships into Syria. Bulldozers and earthmoving equipment dug barriers and assembled protective berms. HESCO Mil concertainer units— gabions made of collapsible wire mesh with a heavy-duty line that were essential force protection tools for allied positions in both Afghanistan and Iraq, and capable of absorbing a punishment of small arms and artillery fire—were positioned around the base. The blast walls were reinforced by additional ramparts, along with shelters and bunkers. US Marine counterterrorist specialists were choppered in to maintain the perimeter and fend off swarm attacks, mortar barrages and vehicle-borne IEDs. Their mission was to protect the intelligence assets who set up shop inside one of the smaller buildings. Machine gun positions controlled the entire perimeter.

Electrical and communications technicians wired the location; generators made sure that the power was never cut off from the outside. Water and food rations were flown in; MREs were augmented by the occasional basket of tomatoes and plastic bags of three-day-old flatbread. Buildings were made livable somehow; air-conditioning units were brought in in preparation of the spring and then summer's inhuman heat. The camp eventually boasted secure WiFi.

Some of the Green Berets called the outpost "the Alamo." Many didn't like the nickname's ominous overtones.

Al-Tanf would have made a Toyota sales representative very proud. The parking lot was full of Toyota pickups modified to sport heavy machine guns—the mechanized "Cadillac of choice" for all of the anti-Assad forces. The Toyotas were all white; some were so new that they looked as if they had been driven off a dealer's lot a few hours earlier. Graphic artists used Sharpies and paint to adorn the doors with the faction logo; a symbol was painted on the cabin roof.

Rebels from across the political spectrum flocked to al-Tanf to receive modern combat equipment and instruction. Training the Syrian forces fell upon the shoulders of Green Berets from the US Army 5th Special Forces Group. Training missions were the specialty of the US Army Special Forces. The 5th Special Forces Group (Airborne), nicknamed "the Legion," was based in Fort Campbell, Kentucky, and worked CENTCOM's sphere of influence:* the Middle East, the Persian Gulf, the "Stans" of Central Asia, and the Horn of Africa. The Green Berets were experts in irregular warfare and training indigenous forces to fight. Operators in 5th Group, because of their responsibilities in the Middle East, underwent Arabic, Farsi and Kurdish language training. The Green Berets were often referred to as ambassadors with camouflaged faces. They were highly skilled in turning ragtag militias into highly trained combat formations.

The most basic Green Beret unit was the twelve-man Special Forces Operational Detachment-Alpha, or ODA. The twelve-man team consisted of a commander, usually a captain; an assistant commander; an operations sergeant; an assistant operations and intelligence sergeant; two weapons sergeants; two communications sergeants; two engineering sergeants; and two medical NCOs. The twelve men were equipped and trained to live off the land, turn inhospitable terrain into shelter and to fight as if the force was four times as large. The Green Berets were highly accomplished shooters and marksmen with their familiar complement of weapons and qualified to use just about any assault rifle, machine gun or antitank weapon in an enemy's arsenal. Most important, to be a Green Beret meant to be an instructor, leaving behind cadres of combat-ready troops that could carry on the fight without American involvement.

Overseas training assignments were traditionally known as

---

* The 1st Special Forces Group (Airborne), based in Fort Lewis, Washington, is responsible for Asia, though it has also seen combat operations in Afghanistan, Iraq and Syria; its motto is "First in Asia." The 3rd Special Forces Group (Airborne), based in Fort Bragg, North Carolina, is responsible for the African continent, though it has seen operations in the Caribbean and in Syria, Iraq and Afghanistan; its motto is "De Oppresso Liber." The 7th Special Forces Group (Airborne), based at Eglin Air Force Base, in Florida, is responsible for operations inside the Americas, including the Caribbean; it has also seen operations in the global war on terror; and the 10th Special Forces Group (Airborne), based out of Fort Carson, Colorado, is responsible for Europe, though it, too, has seen extensive action in the Middle East; its motto is "The Best." There are also two National Guard groups, the 19th and 20th. The Special Forces Groups fall under the US Army Special Operations Command (USASOC), which is part of the overall US Special Operations Command (USSOCOM).

FIDs, for foreign internal defense. It was the bread and butter of the Green Beret list of services, and a twelve-man ODA was dispatched to al-Tanf to begin work immediately. The fear at coalition headquarters, though, was that the allies would end up training ISIS fighters who were masquerading as nonfundamentalist militiamen; there was fear that hostile elements would be able to mingle their way into the secure facility. The rebels had to be vetted carefully and by those who understood their language.

Garrison al-Tanf was divided into two unequal parts. The larger section of the camp belonged to the Syrian opposition forces and it consisted of an eating area, barracks and secure arsenals; the American side consisted of barracks, a mess, classrooms and a technological section for UAVs, communications and other high-tech gear. Each and every Syrian fighter that entered al-Tanf was vetted to verify his background and allegiance, and only those determined to be the genuine article, not undercover members of the Islamic State, al-Qaeda or the al-Nusra Front, were permitted entry and subsequently trained by the Green Berets.* The Syrians were run through a gauntlet of proper firearms proficiency drills because, as one Green Beret officer explained, "their tactics were 'gangster' instead of Range 37," a reference to one of the tactical shoot houses used by the Special Forces groups at Fort Bragg.[254]

Each twelve-man 5th Group team was complemented by a twelve-man Jordanian ODA from the 101st. Although the background of the Jordanian military was British-centric—

---

* An arrangement similar to the Jordanian partnership with the Green Berets took place in a position somewhere between Raqqa and the frontiers of Turkey and Syria, between Jordanian and French special forces battling ISIS. Throughout Iraq, as well, coalition special operations units—from Denmark, the Netherlands, Portugal, Norway, Canada, Australia, the United Kingdom, the United States and even from smaller countries like Latvia—trained the Iraqi police and military, especially the command cadre, in how to lead an offensive to retake territory seized by the Islamic State.

Lieutenant General John Bagot Glubb, or "Glubb Pasha," having commanded the legendary Arab Legion in the 1948 war with
Israel—the military took on a decidedly American influence
over time. The 101st Special Forces Battalion maintained close
ties with 5th Group during peaceful times and, of course, during the redacted moments of covert conflict; the units trained
together, both in the United States and in Jordan. The units had
a history of joint operations beyond the frontiers of Jordan and
Syria. Operators worked together in Afghanistan, where the
Jordanian soldiers were highly successful in the difficult work
of battling an insurgency while winning the hearts and minds
of the local population.

There were no skill gaps between the American and Jordanian operators—it took seventeen months of intense combat,
communications, medical and parachute training for a recruit
to sport the crimson beret of the battalion. But what separated
the American and Jordanian operators was language. Even those
5th Group operators who studied Arabic and spent enough time
in the region fighting wars and training armies could not compete with the Jordanians when it came to the Jordanians' native-
tongue knowledge of Arabic and its regional and religious-based
slang. Joint unit operations became common because of the
ability of the Jordanians—both operators and intelligence officers—to communicate with the local population and the combatants; the nuanced knowledge of Arabic was critical when
debriefing an ally or when interrogating a prisoner. "Our mission with the Americans was to assist, advise, and accompany,"
Colonel Saleem,* a veteran of operations in Afghanistan and
Syria with the Americans explained.[255]

Bedouin intimacy with the terrain and the people of the area
was critical for the allied effort. On one nighttime intelligence-
gathering mission, the American and Jordanian operators were
flown to an encampment north of al-Tanf where it was believed

---

* A pseudonym.

senior Islamic State officers were spending the night. The operators carried extra rations of ammunition and water, in case the engagement turned into a lengthy battle; team members carried flex-cuffs and blindfolds in case there was an opportunity to take prisoners.

The Black Hawks dropped the force an hour's march from the target. It was a moonless night, and moving through the pitch-black was slow and cumbersome, but it ensured the element of surprise. But when the operators arrived, they found the campsite abandoned. Tire tracks were embedded into the cold nighttime sand; the area was strewn with garbage. One of the American intelligence officers, disappointed by the results, told his Jordanian counterparts, "Looks like this is nothing but an abandoned Bedouins camp." But the Jordanian smiled with the confident grin of a man who was about to school an expert. He pointed his flashlight under a hastily assembled blue tarp and pointed to empty plastic water bottles. "The Bedouin don't drink bottled water," the Jordanian responded. "They drink water from their wells." Looking around, there were no goat droppings, no telltale signs of animals of any kind. Food wrappers were found, as were a few half-eaten sandwiches. There was human waste near the tarp where, it was evident, people slept. "The Bedouin don't travel without their animals or kids," the Jordanian advised. "And they don't shit where they eat and certainly not where they sleep. They respect themselves and the desert too much for that."[256]

Instead of calling in the helicopters for the extraction, the search continued in the desert until several Islamic State operatives were cornered and killed. Such missions were common.

There were attempts by Islamic State units to cross the Damascus–Baghdad highway near al-Tanf by masquerading as local Bedouins. But there were precise aspects of how the Bedouin spoke, acted, walked and interacted with others that were hard to imitate and easy to detect by the experienced hands of

101st Battalion and GID personnel. Foreign fighters, especially men who spoke French or German better than Arabic, or the men from the Caucasus, who were blond with blue eyes, made gallant but doomed efforts to fool the Jordanians. "We could tell just by looking at someone or by hearing him speak if they were a *bedawi*, a Bedouin, a *falah*, or farmer, or an *al-medani*, a city dweller," a GID officer explained. "How could someone who spent their entire lives in the United States or Great Britain know this? This knowledge wasn't something you could pick up in a course. It required a lifetime of living in the area, living with the various groups of people, and even hearing tales spoken at family gatherings, to be able to know someone's history before they've had a chance to tell it to you."[257]

Islamic State operatives who were apprehended at these checkpoints or ambushes weren't all the tough-as-nails fanatics that were advertised in al-Adnani's propaganda. They were men, mere youngsters in some cases, who didn't want to die for a cause they saw as being on the losing side; it was a sign to the CIA and GID intelligence officers, at least, that the war had changed decidedly in the coalition's favor.

The prisoners were eager to cooperate, especially for some water and food, and a promise that their lives would be spared. The intelligence they provided was not only current and reliable but intimately revealing of the many political and ethnic internal conflicts underway inside the ranks of the Caliphate.

Garrison al-Tanf became a highly effective launching pad for proactive raids against Islamic State convoys and fighters crisscrossing the area. The Islamic State soldiers rarely traveled alone. They moved with their wives and children in tow. Some of the brides were nothing more than children, sex slaves seized somewhere along the front lines for perverse pleasure. Often these teenage brides were teenage mothers, and when the Islamic State moved, it did so with small children and infants riding inside the sardine-like crunch of an overpacked sedan. The wives and children were driven in cars that were packed with explosives. "The families were the ideal human shields," a GID officer explained. "The khawarij were cowards, and they knew that we didn't fire upon women and children."[258] The Green Berets and the 101st Special Forces teams were always careful when they encountered families in the caravans because they could be refugees trying to reach Rukban, or foot soldiers of the Caliphate moving from one location to another.

The loss of al-Tanf was a serious blow to the Islamic State coffers. Narcotics and oil smuggling were significant sources of Caliphate revenue, and without free access to the main Baghdad–Damascus highway, commerce was interrupted, and the leadership's income put in jeopardy. The business of terror was racketeering, and al-Tanf denied the clerics in Raqqa the monies needed to continue and expand the war. Abu Omar al-Shishani ordered al-Tanf retaken. It became an obsession for him, especially in the ruling council where, as one from the Caucasus, he was never on truly equal footing with the Iraqis and Syrians who were really in charge of the day-to-day operations. Al-Shishani's instructions to his legions in the field were specific and immediate: retake al-Tanf. A special reward was promised to anyone who captured an American or Jordanian.

Garrison al-Tanf absorbed occasional mortar and artillery fire, and swarm assaults by bands of men attacking from all sides, but with air support, UAVs, snipers, and machine gun and mortar positions, Islamic State efforts always fell short. Al-Shishani ordered his workshops to make vehicle bombs (VBIEDs) that could crash through barriers and withstand fire—these rigged trucks were fitted with plates of steel around the engine block and the driver's compartment to keep the suicide bomber alive as he prepared to drive into a target. The allied arsenal fielded many weapons, such as the FG-148 Javelin antitank missile, that could destroy these truck bombs from safe distances.

In the field against conventional formations—especially allied special forces—the Islamic State was not the fierce army that it seemed to be when it seized Mosul in 2014.

Wherever the CIA and GID established a presence, Russian and Iranian intelligence paid close attention. The Russians were initially quite pleased—even amused—that coalition forces and the Islamic State were battling and killing each other in fire-fights and air strikes that, ultimately, benefited the survival of the Assad regime and killed militants, especially those from Chechnya and the other former central Asian republics, who otherwise could one day attack targets in Moscow and Saint Petersburg. But al-Tanf worried Russian commanders in Syria. The more men and matériel that flowed into the base, the Russian generals concluded, the more permanent it would become, and that upended President Putin's orders to turn back the clock and restore Bashar al-Assad to power in *all* of Syria. Because al-Tanf was so close to Iraq and Jordan, the Kremlin feared that its strategic proximity would become an argument used in international forums to maintain an American-led bastion inside the country.

The Russians realized that any attack on the garrison at al-Tanf had the potential to dangerously escalate the anti–Islamic State effort into a much wider conflict—so they did the next best thing: dangerously provocative harassment. The GRU, with its elite Spetsnaz special operations force, constantly stalked al-Tanf, monitoring the comings and goings of American and Jordanian helicopters, along with the types of vehicles and armed men that moved in and out of the camp. The Iranians' Islamic Revolutionary Guard Corps, as well as the more sinister Quds force and other directories from the MOIS, were also involved in the area's thriving narcotics trade—traffic moved in both directions on the Damascus–Baghdad highway and the poppy fields of the Bekaa Valley in Lebanon and western Syria constituted a large percentage of the income expected by their bosses in Tehran.

With so many spies and soldiers moving around in a fairly

confined area, the American and Jordanian contingents were restricted by very strict rules of engagement that demanded caution or even overcaution. In some cases a coalition officer recalled, "It was a cat-and-mouse game with enormously high stakes. We were armed, they [other intelligence services] were armed, and the chance for a mistaken shot to become a full-fledged firefight was always there."[259] There were always eyes on al-Tanf and, the operators garrisoned inside the fortified gates guessed, sniper scopes and targeting orders, as well.

In order to add a layer of deniability to any possible encounter, the Russians also employed mercenaries from the Wagner Group. In Syria, the Wagner Group was used to train indigenous forces, even Assad's elite commandos, to carry out a scorched-earth policy against the Islamic State and the towns and villages that harbored them. The Wagner Group was private, they were mercenaries, and, as a result, they provided Moscow with plausible deniability in operations around the world and particularly inside Syria that served Russian interests.[260]

Mostly, though, the Wagner Group guns for hire were paid to watch and monitor allied activity and avoid a shootout with American forces. If there was an incident, the Kremlin surmised, they could always distance themselves from those killed by claiming they were private guards sent to Syria to guard oil installations and not combat soldiers acting at the behest of the Russian government.

The Russians wanted to avoid gunfire, especially as the presidential election cycle in the United States was turned inside out by Donald Trump's race to the White House. So Russian intelligence restricted its harassment of Garrison al-Tanf to electronic eavesdropping, radio jamming, and interfering with GPS signals in the choppers that flew in and the drones that were dispatched on reconnaissance sorties.

CENTCOM HQ, al-Udeid Air Base and even Amman were far removed from the remoteness of the al-Tanf forward operating base. Summer arrived at al-Tanf like it always does—with roasting winds and a merciless sun. The Marines were used to the harsh conditions, especially those who had served in Iraq before. They stood guard and protected the perimeter, while the operators and spies inside the base worked on operations and closing the noose on the Islamic State leadership.

Amman was far removed from the white-hot dust bowl of al-Tanf. King Abdullah heard daily progress reports concerning the coalition's progress against the Islamic State. He received updates concerning the air force's bombing campaign on behalf of the Inherent Resolve alliance, and situation reports from his special operations commanders concerning the situation at al-Tanf. The GID also provided a brief to the king with the most pressing news from the region, as well as how the covert fight against the Caliphate was going. The Moaz Task Force report was an essential piece of the king's daily brief. The king realized that accurate intelligence, like planning for a commando assault, couldn't be rushed. The reports, though, were promising. Assets in the field were being pressed. No stone was left unturned.

17

Just Rewards

The desert does not mean the absence of men, it means the presence of God.

—Carlo Carretto[261]

IN THE SPRING of 2016, the men of Abu Bakr al-Baghdadi's once unified ruling council found themselves in the grips of squabbles and backstabbing. Gone were the days when these holy men—the self-anointed clerics and the killers who wrapped themselves in their interpretations of the Koran—agreed with whatever the caliph demanded. These men might have been fanatics, but they weren't fools: battle hardened and pragmatic, they found it impossible to whitewash the reports from the battlefields in Syria and Iraq that detailed the worsening military situation for the Islamic State and its military assets. Most important, the ledger book was showing that oil, smuggling antiquities, extortion and sexual slavery profits were down. Religion and rage built the enterprise, the ruling council knew, but money was what made it function.

The jihadist army that had raced across the Middle East only two years earlier and spread fear into hearts and minds around the world was now stalled and fighting to hold on to every inch of territory it had seized. The combined tonnage of Russian and coalition ordnance was too much for frontline Islamic State units to absorb; the round-the-clock bombing resulted in losses on the battlefield. Every town and village that was recaptured in Iraq or Syria, every portrait of President Assad or banner from the YPG, the Kurdish People's Protection Units, that replaced the black banner of the Caliphate inside the local police station spelled out the unmistakable reality for al-Baghdadi's council and their generals in the field. The ambitious campaign to conquer the Levant and then the entire Middle East in some fantastic medieval adventure was ultimately no match to withstand the conventional might of the world's top military powers.

The Islamic State needed—to borrow from General David Petraeus's playbook—a surge: a mighty multifront military gesture to redefine the war and how it was going to be fought. Abu Mohammed al-Adnani, fresh off his game-changing victories in Paris and Brussels, would expand global operations. Abu Omar al-Shishani was responsible for reinvigorating the fervor his men had summoned in Mosul and elsewhere to hold the line until attacks overseas could attract a new army of jihadis. All eyes turned to Ash Shadaddi, a town in northern Syria near the Iraqi frontier, and one of the pins on a map that the Islamic State was determined to somehow hold.

Ash Shadaddi was one of the first towns in northeastern Syria to fall under the black banner in 2013. Some fifteen thousand residents called it home before the Islamic State took over—there is no count as to how many were killed by the Emni and the sharia enforcers. Store owners who paid the Caliphate tax had their shutters painted with Islamic slogans; those who refused to pay were usually tortured inside the town's police station and then hanged from an electric pole. Women were executed for showing too much flesh under their hijabs or for smoking in public. Punitive killings were not new in the town: the Turks, the French and Assad's henchmen always disposed of those charged with minor offenses in the past. With the exception of electrical lines and satellite dishes, Ash Shadaddi hadn't changed in a thousand years. It had a single dirt road that sliced through the center of town, along which donkeys pulled carts that brought the morning's harvest to market.

But Ash Shadaddi was one of the last towns near the Iraqi frontier that stood in the way of Raqqa. It was near oil fields and trucking routes to Turkey. Truckloads of rockets and artillery shells were rushed to Ash Shadaddi, as were reinforcements from other sectors. In October 2015, though, Kurdish forces and other smaller allied Sunni tribal militias pushed against Islamic State bastions in Syria's north, including the area around Ash Shadaddi in what was known as al-Hasakah Governorate. When Islamic State fighters began sending their family members south, it was an indication that a bitterly contested fight-to-the-last-man battle was to come. The town fell after bitter fighting.

The loss of the town infuriated Abu Bakr al-Baghdadi. He personally ordered Abu Omar al-Shishani to the front lines to bolster his troops. The redheaded Georgian was never considered a brilliant battlefield tactician. Rather he was viewed as an inspirational leader; his talent was raising the morale of the men in the field, primarily the Caucasians who revered him and fought like fanatics on his behalf. Al-Shishani's tour of the front lines was designed to spark a fire into the hearts of the men who were fighting on a daily basis against air strikes and conventional ground assaults. Al-Shishani was a brilliant motivational speaker, and the men—particularly the foreign fighters from Europe—needed a reminder as to the religious significance of their sacrifice.

Al-Shishani's itinerary was a closely guarded Islamic State secret, but in fact coalition forces were certain they knew exactly where he'd be and when.

In early 2016 the United States initiated a special operation offensive against Islamic State targets in Iraq and Syria. The force, known as the Expeditionary Targeting Force, or ETF,[262] consisted of the top-tier of JSOC and CIA special operations units that went after high-risk and top-level Islamic State commanders. The ETF's primary mission was to capture the most important men in the Caliphate's leadership—and to seize their phones, laptops and notebooks—for intelligence purposes. The ETF consisted of close to two hundred men and included safe houses and Iraqi and Syrian Kurdish intelligence officers who could facilitate interrogations and prisoner warehousing, so that new versions of the notorious detention facilities, such as Abu Ghraib prison, wouldn't have to be opened.[263] Reports indicated that the majority of the personnel assigned to this top secret group came from Delta Force and that they had captured a key Islamic State officer that helped the NSA and the CIA to cut into al-Shishani's direct line of communication.

The ETF's mission was virtually identical to other specialized hunting teams that JSOC had set up before in the Middle East. Task Force 121 was the unit established in the wake of Operation Iraqi Freedom to capture—or kill—Saddam Hussein and his sons Uday and Qusay;[264] Task Force 6-26 included virtually every direct-action and aviation asset of JSOC, US-SOCOM, the CIA, the NSA, the DIA and even the FBI, and went after Zarqawi's network in Iraq in 2006.[265] British special operations teams from the Special Air Service and Special Boat Service (SBS) also participated in this multinational and multiagency manhunt.[266]

Old-school masters of the human intelligence game would characterize the methods of such task forces as impatient, hurried. Harsh interrogation techniques were frequently employed on the men—and women—snared in task force raids. Human rights were abused, and individuals were tortured in hastily established holding centers that, in the case of Task Force 6-26, ultimately produced more future recruits to the Islamic State than actionable intelligence on Zarqawi's whereabouts. The task force reported that its unofficial motto was "If you don't make them bleed, they can't prosecute for it."[267] In an attempt to learn from their previous errors, the ETF focused on the peripheral intelligence that could be gained by unlocking cell-phone communications and breaking into encrypted laptops. The ELINT and SIGINT intercepts were the sort of scandal-free intelligence that was always preferred by the decision makers in the Beltway and, as a result, by the colonels and generals at CJTF-OIR headquarters, as well.

On March 4, 2016, coalition aircraft were launched to strike at a building near Ash Shadaddi where this electronic intelligence pointed to the presence of Abu Omar al-Shishani; the redheaded Georgian was in town giving a pep talk to his senior lieutenants. The intelligence was never the on-the-ground, eyes-on-target variety but rather high-resolution imagery from drones, reconnaissance aircraft or satellites. The bombing run's yield was as wide as an apartment block and several meters deep. No one, the post-target evaluators assessed, could have survived.

At first the coalition was cagey in its will to discuss Abu Omar al-Shishani's departure from the Islamic State's roll call. In a matter-of-fact press release, coalition spokesmen dispatched word that aerial assets had conducted four separate strikes in the confines of the al-Hasakah Governorate, which hit three separate Islamic State tactical units and, in the process, destroyed a machine gun position, an Islamic State vehicle and an Islamic State tunnel system.[268] There was no statement from the Raqqa Media Center, either. Al-Baghdadi's spokesmen were usually quick to confirm when one of their senior men was dead. There was no mention of al-Shishani—no definitive word of his death or even a hint that he escaped harm. The Pentagon, however, was impatient to make headlines with al-Shishani's death. On March 9, Pentagon press secretary Peter Cook gathered media members who covered the Department of Defense and announced that Abu Omar al-Shishani had, indeed, been killed in a coalition air strike. "Batirashvili is a battle-tested leader with experience who had led ISIL fighters in numerous engagements in Iraq and Syria," Cook said. "His potential removal from the battlefield would negatively impact ISIL's ability to recruit foreign fighters, especially those from Chechnya and the Caucasus regions, and degrade ISIL's ability to coordinate attacks and defense of its strongholds like Raqqa, Syria, and Mosul, Iraq."[269]

On March 15, news of al-Shishani's elimination was confirmed by the *New York Times*. "A senior Islamic State militant whom the United States military tried to kill in an air strike in Syria last week has died of his wounds, according to a senior Pentagon official," the report announced.[270] US Defense secretary Ash Carter was gleeful in this latest victory of the Islamic State's leadership.

The only problem was that Abu Omar al-Shishani wasn't dead.

There was a standing policy inside the CTD in particular *never* to travel to downtown Amman to buy celebratory sweets solely on a rumor. Rumors were the death knell for intelligence agencies who dealt in inside knowledge and never traded in speculation. The GID knew that al-Shishani was alive because, unbeknownst to the CIA and the coalition, the GID had nurtured a source inside the Islamic State that had access to al-Shishani. For operational security considerations, his identity was and remains top secret, as is the nature of the actual relationship; information concerning his name, age, background, past affiliations and previous activities remain closely held to the vest and was not and could not be shared with the American Friends. All that the GID could confirm was that he was able to interact with the Islamic State minister of war and that he wasn't an Arab. The source, known by a code name that we will call "Victor,"* had risen through the ranks of the foreign fighters and was considered a trustworthy and loyal individual dedicated to the struggle against the West. He was a man who displayed unique courage on the battlefield; his pedigree and his CV were above reproach to even the suspicious thugs of the Emni.

Victor was a dedicated jihadist, someone who believed in the struggle, yet somehow† the GID managed to gain Victor's trust and secure his cooperation. It is not believed that the lure was financial, because the Americans usually bankrolled such expensive contacts and, in return, demanded some level of reach in controlling the asset. The reasons behind Victor's decision appear to have been personal; though they didn't involve sex or religion.[271]

Deep-planted sources were, as one of the Jordanian intelligence officers explained, "prized jewels that could only be brought out on special occasions. They weren't in play for flash or to impress a friend. They were never to be shared. They required safekeeping and they had to be protected at all and at any cost."[272] Victor's true identity was known only to his handler and the handler's immediate commander. References to his true name and place of birth were never committed to paper. His presence inside the ranks of the Islamic State was never shared with the Americans. To safeguard his identity, the Jordanians were careful not to counter the coalition's claim that al-Shishani had been killed in March.

Al-Shishani was Victor's primary mission. Like any deep-cover agent, he was ordered to risk communications only to

---

*   Not the actual code name.

†   The exact means and measures of the recruitment process remain highly classified.

relay vital information. "Islamic State chatter was beneath the services of such a valued source," a retired GID officer explained. "Such a valued possession required patience and perseverance. And, most of all, protection."[273]

The GID saw no need to risk exposing Victor to the Emni in order to obtain verification that al-Shishani had been killed in March. Firstly, the Islamic State was a great many bad things, but they made a concerted effort to disseminate truth; they didn't make grandiose claims, nor did they deny attacks that they perpetrated. "The organization tried to spin a narrative that what they said about themselves is true," Middle East expert Firas Abi Ali commented. "Though one could never really know how their bureaucracy worked, except for what they say about it and from their documents."[274] Victor's case officers knew that he would make contact if and only when there was such earth-shattering information that the risk was worth the reward. The opportunity that the GID had waited for emerged in late June.

The summer in Syria and Iraq was as much a change in the weather as it was an artillery barrage. The heat made every aspect of warfare harder for men who fought on the ground, in the desert, and without the luxury of air-conditioned cockpits and refrigerated meals. The heat bombarded everything and destroyed much. Weapons left in the day's sun became too hot to handle; armored vehicle engines overheated. Men, especially the foreign fighters who were used to the temperate climates of Western Europe and the Caucasus, fell ill to dehydration and dysentery. Clean water was the difference between life and death; transporting water supplies, even when delivered in petrol-laced jerry cans that came close to boiling in the sun, was as important to frontline forces as supplying them with ammunition. Food storage and waste removal became critical issues for men who had scant little training in leading conventional fighting forces. The Bedouins—and Hollywood—thought of the desert as clean. But in reality the vast open roast of a merciless sun cooking sand and dune void of shadow and shelter was a dangerous place where one could meet a painful and dirty death.

The harsh summer conditions were exacerbated by the incessant allied and Russian air strikes. The round-the-clock bombing made it harder for the Islamic State to carry out its daily affairs and to lead the defense of its frontiers. By the summer of 2016, the Caliphate had already lost 25 percent of its landmass. Air strikes and battlefield developments made it hard for the Islamic State's military leadership to plan and coordinate high-level meetings. It was difficult to coordinate general staff field meetings when so many of the men who led the fronts were targeted for extrajudicial termination by coalition planners and Russian generals. It took days, sometimes much longer, to coordinate the top secret sit-downs. The front had been stagnant along a stretch of several key areas in Syria and Iraq, and it was suspected that coalition forces would propel the conventional Iraqi Army in a heavily supported bid to retake Mosul. The Peshmerga were already moving forces and special operations assets—assisted by US Navy SEALs and other coalition commandos—toward staging areas where the Peshmerga was mobilized and ready to lead the campaign to recapture Iraq's second-largest city. The fall of Mosul would, in essence, mean the end of the war.[275] The Mosul theater of operations, from the Islamic State's perspective, had to be protected at all costs.

At the end of June, al-Shishani called such a meeting for his key sector commanders; attendance was mandatory. The date selected for the top secret gathering was July 10. The location picked was the small town of al-Shirqat in eastern Iraq. Information concerning the planning session was highly compartmentalized. Al-Shirqat was situated on the west bank of the Tigris River and was fifty miles due south of Mosul; it was also twenty miles from Makhmur, a Kurdish stronghold in what was known as Sector 6. Al-Shirqat was a city with a vast martial history.

It was built near the ancient Assyrian fortress at Assur, and in 1918, British and Ottoman forces fought one of the last battles of the First World War for control of the city.

Al-Shishani's presence at any high-level strategy session was, if nothing else, a reassuring sign for the lieutenants and field commanders who would lead the defensive actions and counterattacks. Men, especially volunteers from the former Soviet Asian republics, would eagerly volunteer for suicide bombing strikes if they knew that their beloved "Omar the Chechen" had directed them to do so. Al-Shishani maintained a powerful loyalty inside the ranks of the Inghemasiyoun, the Islamic State special forces. They would be deployed in and around any advance on Mosul to wreak havoc inside the ranks of Iraqi Army conventional forces marching toward the city; once their presence became known to the underpaid and poorly motivated Iraqi infantrymen, any allied attack would collapse even before it began. Their war cry of "victory or martyrdom" terrorized their opponents on the battlefield.[276]

High-level gatherings came with strict security protocols, as Emni agents camped out in the area to make sure that possible venues weren't under surveillance from the air or from coalition agents already on the ground. Local residents were interrogated; their homes were often seized so that the security specialists could create a discreet yet protective perimeter.

The location ultimately selected was a large home that was slightly away from the center of town and close to the banks of the Tigris River. Isolated was ideal. When the GID recruited Victor, his case officer made it emphatically clear to the asset that the coalition could not use any information concerning targeted individuals that would ultimately result in the deaths of civilians. Victor's case officer gave him an emphatic warning, a CTD team leader code-named Abu H.,* remembered. It was "No kids, no women, no innocent people."[277]

---

* A pseudonym.

Due to security considerations, Victor couldn't simply text the details of the place and time. Emni agents routinely confiscated all mobile devices—tablets, mobile phones and laptops—belonging to individuals close to an inner circle who could cause severe damage with the information they came in contact with. Victor would have to transmit the news via age-old tradecraft: coded messages on paper left in dead letter drops. There were contingencies in place for such archaic methods, when safeguarding the communications and the messenger was vital.

The exact makeup of Victor's message is classified, as are details of precisely how it was delivered to his case officer. What can be told is that the coordinates of the meeting, the target for the coalition, arrived as a screenshot from Google Maps.

The message's arrival sparked a flurry of activity in GID headquarters. The information in the message had to be verified as genuine, as did the message itself—a fear was that Victor had been captured and the material now in Amman under review was a trick by the Emni to lure coalition forces or aircraft into an ambush. Once the information was deemed as beyond suspect, Victor's extraction had to be worked out, and the details of his coded broadcast disseminated to Amman Station. The CIA was better equipped than the GID to rush the news and location of al-Shishani's war council meeting in al-Shirqat to coalition intelligence and targeting officers so that they could plan the multiple details and contingencies needed when launching aircraft over enemy territory. "There was nothing in this process that was instantaneous," commented Rick,* a US Special Operations Command officer who worked the front lines with Syria. "There were protocols and commands that had to be brought into the loop."[278]

The GID provided their liaison colleagues in Amman Station a neat and highly sanitized report of al-Shishani's location. The dossier went high up the chain of command, undergoing redac-

---

*   A pseudonym.

tions and alterations until all that the pages reflected were the type of source, the information, the target and the coordinates. Two CIA liaison officers came to GID headquarters to pick up the envelope. The GID wondered if their superiors in Amman Station or Langley would be wary of acting on the intelligence that countered their narrative and the triumphant press conference.

Sixteen top Islamic State field commanders greeted Abu Omar al-Shishani as he walked down a darkened staircase in the cellar of a two-story home that was large enough for at least three families.[279] The men looked tired, but they became energized to see al-Shishani and his phalanx of bodyguards. A highly detailed map was spread across a large dining room table. Smaller maps were spread across chairs and draped on the wall. Small fluorescent light bulbs illuminated the room. The air inside the room was stifling with cigarette smoke. A fan buzzed as it twirled a steady rotation and blew a slight breeze of air inside a room that was crowded and stifling. The lieutenants parked their Toyota Land Cruisers far away from the house so that satellites and drones wouldn't be able to pinpoint exactly where the high-level gathering was taking place.

The American combat aircraft approached from the east, from the rising sun, and through the established corridor of attack against the Islamic State. The aircraft snuck in completely by surprise and were not subjected to any antiaircraft fire. The 500-pound precision bombs that were launched against the building hit at dawn's first light. The destruction of the two-story structure was absolute—all that was left of the building was a forty-five-meter-deep smoldering crater. No one survived the attack.

The following morning, the CJTF-OIR public affairs machine released word concerning the previous day's aerial activity against the Islamic State and two strikes near Mosul, claiming that five Islamic State assembly areas had been hit.[280] There were no mentions of al-Shishani. The next day, though, the Syrian-

based Amaq News Agency, a service with direct ties to the Islamic State, released the news that al-Shishani had, indeed, been
killed in the coalition action; the communiqué was dispatched
in Arabic, English and a multitude of other languages, including Russian and Bengali. The Islamic State was adamant about
disseminating an accurate narrative about its forces. Amaq's dis-
patch that al-Shishani had been killed was as good as an official
confirmation.

To the outside world, the coalition company line was one of
caution. US Army Lieutenant General Sean MacFarland, the
commander of the Inherent Resolve coalition, expressed his
confidence in the intelligence that targeted al-Shishani, though
he was reticent in formally declaring the Georgian dead. "We're
being a little conservative in calling the ball on whether or not
he's actually dead or not. But we certainly gave it our best shot,"
MacFarland explained to reporters covering the war in the Iraqi
capital. Behind the scenes, far from the prying eyes and digi-
tal recorders of the international press, the overall sentiment at
CJTF-OIR headquarters was one of relief. Declaring one of the
top objectives of the military campaign dead months before he
was actually killed had been something of an embarrassment
for the coalition.

Word of al-Shishani's death was confirmed by Victor. No one
removed a bottle of aged scotch from a desk drawer at GID headquarters; alcohol wasn't consumed on the compound, after all,
but now *was* the time for several of the counterterrorist officers
to stop at one of the five Amman branches of Al-Haj Mahmoud
Habibah & Sons Company, a famed pastry chain that originated
in Jerusalem, for boxes of sickly-sweet cakes and fried treats to
hand out throughout the office.

The removal of al-Shishani was an enormous accomplishment for the GID and its unrivaled HUMINT capabilities. Al-
Shishani had been a top-priority target for the coalition for

close to two years; enormous resources were invested in orbiting satellites, launching drones and paying off informants for any information on his whereabouts. The might of the Russian military—and the ruthless hand of their highly effective intelligence apparatus—left no stone unturned in the hunt for the Georgian who motivated the Chechens and the Tajiks to wage a holy war against Moscow. Ultimately, though, it was the well-placed asset of the small Jordanian service that achieved the desired results.

The termination of Abu Omar al-Shishani opened a release valve for the pressure-cooker environment that had developed inside the CTD. The HUMINT penetration of the Islamic State inner circle was a masterful achievement, and the men who handled the asset and the information were glad that their efforts yielded such high-value and actionable intelligence. Most important, the GID director was very proud of his officers, as was King Abdullah. The Jordanian monarch read the full report of the operation that led to the July 10 air strike, and he was pleased with the effort made by the Moaz Task Force. The use of an invaluable source as fuel for a multinational effort personified Jordan's role in the international war against the Islamic State and other transnational players. Most important, al-Shishani's death was a measure of payback, and in the Middle East revenge mattered.

There were four more names on the list: al-Adnani, Omar Mahdi Zaidan, Abu Khattab al-Rawi and Abdul-Hai Al-Muhajir. There was a lot more work that needed to get done, though. But there were also several more men like Victor out there behind enemy lines.

18

Find, Fix and Finish

When justice is done, it brings joy to the righteous but terror to evildoers.

—*Proverbs 21:15*

EACH MORNING, AT the start of their workday, the intelligence officers assigned to the Moaz Task Force made a point of examining the few photos of Abu Mohammed al-Adnani that they possessed. The images, some in black-and-white, were either printed off video feeds or obtained from sources in the region. The photos were pinned to a whiteboard and marked with small notations scribbled on Post-it notes. Al-Adnani had natural good looks, a prerequisite for a man who would attract followers to a cause. The combination of charisma and a homicidal disposition made him someone to be feared; it was said that al-Adnani had an explosive temper and was known to exhibit daylong fits of rage. But there was something else about the man that drove the GID hunters mad. He displayed an arrogant come-and-get-me smirk that looked as if he was daring the spymasters to see if they were smart enough to find and kill him. The intelligence officers gazed at the images of the man in an attempt to figure him out, but they couldn't. He remained a frustrating question mark.

The men on the Moaz Task Force read and reread the transcripts of Abu Mohammed al-Adnani's speeches. Some of these ramblings were bizarre and depicted a man saying crazy things for the sake of appealing to an audience that wanted to hear a diatribe, such as when he said of President Obama, "O Obama. O Mule of the Jews, you are vile."[281] They listened to intercepts of his conversations and heard the cadence of his voice and the language he used. They wanted to psychologically dissect al-Adnani from every angle and construct a forensic road map cataloging where, when and how al-Adnani was turned from a religious young man into a monster who could choreograph cold-blooded murders for the sake of publicity.

The men of the Moaz Task Force were especially interested in al-Adnani's upbringing, hoping that if they could find a vulnerable landmark in his life—human indicators, such as friends or teachers who were influential in his life—that were critical to building a matrix of contacts and locations that could be used to find where he was hiding. Their work began in the town of Binnish, in the northwestern corner of Syria quite close to the Turkish frontier, where Taha Sobhi Falaha, the man that would ultimately be known as Abu Mohammed al-Adnani, was raised and transformed into a militant.

Binnish wasn't downtrodden and impoverished like the jihadist breeding grounds around Deir ez-Zor and Iraq's Anbar Province. The town dated back two millennia before Christ and was well-known throughout the region for its succulent olives and sugary figs. During the days of the Mamluks and the Ottomans, Binnish was a resort area. The clean mountain air and majestic landscape attracted travelers from around the empire. The French colonial masters vacationed in Binnish to flee the urban blight and political skullduggery of Damascus. Binnish was a religious town with a grand mosque built in the fifteenth century; it was also an intersection of tolerance where different ethnicities and religions of the region intermingled without animus or sectarian violence.

President Hafez al-Assad's regime changed the Syria of old forever, especially towns like Binnish where the Baathists ruled without mercy. They arrested school headmasters and professionals, almost exclusively Sunnis, who didn't toe the party line. Taha grew up in a religious household. The men at the mosque, his teachers, were his role models. It was claimed that, as a young boy, Taha memorized the entire Koran in less than a year.[282]

The Assad regime was always wary of the pious, and the religious men were singled out for special treatment—especially clerics who were part of underground political movements and religious orders. Public executions were routine. Taha was five years old when President Assad's tanks destroyed the city of Hama, fifty miles south of Binnish, in February 1982, following an uprising by the Muslim Brotherhood. Assad's shock troops killed thirty-eight thousand people in a twenty-seven-day siege in what would become a harbinger of things to come.[283] Refugees fleeing Hama told tales of unspeakable atrocities and of young men taking up arms to defiantly challenge Assad and the Alawites. "Oppression made you quiver in your boots, or it motivated your defiance," Dr. Samir,* a former GID psychologist, explained. "And if this oppression motivated defiance that was religious in nature, the end result was absolutely dangerous."[284]

As a teenager, Taha Sobhi Falaha ignored conventional schooling and spent his days in the mosque with like-minded men listening to the elders talk of those who fought back against the Assads and their Alawite cronies. The mosque was where impressionable young men went to receive an education in how to vanish when the dreaded mukhabarat was in pursuit, and where to hide money and weapons. Taha admired Osama bin Laden

---

* A pseudonym.

and the Afghan Arabs who defeated the Soviets in Afghanistan. He started his *known* jihadist activities in his late teens when he launched a daring campaign to convert Shiites and others. His proselytizing activities brought him to the attention of the Syrian security forces.[285] By his twenty-third birthday, he already possessed a secret-police file that was as thick as a phone book. Ultimately he was imprisoned (trials were not necessary); as a Sunni jihadist, he was tortured, beaten and caged with other jihadists. These prisoners became his extended family—men who proved their trustworthiness as they hung upside down in a cellar torture chamber and had the soles of their feet beaten with a lead pipe.

When the United States invaded Iraq in 2003, President Bashar al-Assad emptied his prisons of the Sunni jihadists and led them across the frontier to fight the American-led coalition. Taha was one of the first volunteers. He was a very different man from the one dragged into a prison cell in northern Syria. He joined Abu Musab al-Zarqawi's army and became known as a fierce fighter who gave no quarter to his enemy.

American forces apprehended Taha in 2005. He carried forged papers in the name of Yaser Khalaf Hussein Nazal al-Rawi;* interrogators believed him to be an Iraqi, even though his Syrian accent should have been a clear indication as to his background.[286] There are no reports that any other allied intelligence service was offered the chance to interview the man from Binnish.

Taha spent six years in the jihadi finishing schools of Camp Bucca and Abu Ghraib. His reputation preceded him in custody and he emerged as a leader of the nascent movement that al-Baghdadi and his lieutenants had hatched behind the barbed-

---

* According to Interpol, other aliases used by Abu Mohammed al-Adnani included: Nasser Khalaf Nazzal Alrawi; Jaber Taha Falah; Abou Khattab; Abou Sadeq Alrawi; Tah al Binchi; Taha Sobhi Falaha; Abu Baker al-Khatab; Abu Sadek al-Rawi; Taha al-Banshi; Abu Mohamed al-Adnani; Abu-Mohammad al-Adnani al-Shami; and Hajj Ibrahim.

wire concertina. By 2011, when the leadership of what would become the Islamic State broke out of their prisons, Taha was sent by al-Baghdadi to Syria to serve as one of the organization's key military commanders in the fight against Assad and in the creation of the Caliphate. He adopted the nom de guerre of Abu Mohammed al-Adnani and became the second most important person—and target—of the organization.

The path from young Taha to the hardened fanatic who ultimately became al-Adnani was a long and bloody sixteen-year journey.

The Moaz Task Force scrutinized every location where they knew al-Adnani had visited. They tried to cross-reference every acquaintance—terrorist and other—that he had crossed paths with. It was a Herculean undertaking made more difficult by battle lines that shifted almost every day.

Abu Mohammed al-Adnani was not a faceless silhouette in a top secret file. He appeared in *Dabiq* and in other media outlets dressed in combat outfits—always in black, with his chest pouches holding magazines for his AK-47 assault rifle. It was essential that he appeared as if he was a frontline warrior—his personal image to the outside world meant everything to him. And always, as one GID officer noted, he posed with his cocky stare. Some assigned to the task force called al-Adnani an "Abu Jasa," a less-than-complimentary slang term for someone who thinks way too much of himself and for whom nothing short of a beating would bring him down a peg.

One of the challenges that the task force faced in trying to pinpoint al-Adnani's location was that he was obsessively paranoid. Al-Adnani trusted no one—from the French-born volunteers he recruited into sleeper cells he would later send back to Europe, to, especially, al-Shishani's legion of Chechens, Georgians, Dagestanis, Turks and Uzbeks; they were all Russian agents, he feared. Remarkably, although his face was featured on State Department Diplomatic Security Service posters and matchbooks throughout the Caliphate, recruits who completed their ten levels of operational training and who met al-Adnani had to do so blindfolded so that they wouldn't know what he looked like.[287] Al-Adnani often moved around the Caliphate in disguise, sometimes even wearing a black abaya and hijab. Emni operatives ran fake motorcades to confuse the intelligence services following him; he scheduled meetings that he had no intention of ever attending. Al-Adnani rarely slept in the same bed twice.

Part of al-Adnani's paranoia stemmed from his belief that he was the most important man in the Islamic State hierarchy and that if he were to be assassinated, the Caliphate's military capabilities would evaporate overnight. In many ways he was right: he led one of the most ruthless counterintelligence units in the history of terrorism, and the incredibly effective Madison Avenue message of barbaric violence was his creation. But al-Adnani was also a psychopathic narcissist who demanded that the narrative always focus on him. It was an odd psychological profile for a man with a $5 million price tag on his head. A GID officer used to joke about al-Adnani, saying, "It's easy to be paranoid when everyone wants to kill you."[288]

The paranoia worked. Al-Adnani disappeared from view while a half dozen intelligence services actively hunted him. As satellites and spies did everything in their power to locate the elusive number two in the Islamic State, al-Adnani continued to plan, plot and attempt to replicate the success of his Paris and Brussels massacres. In May an audio clip of al-Adnani was released to coincide with the holy month of Ramadan. His rant, an effort to spark suicide attacks in Europe and the United States, claimed, "Ramadan, the month of conquest and jihad. Get prepared, be ready to make it a month of calamity everywhere for the non-believers...especially for the fighters and supporters of the caliphate in Europe and America."[289]

There were reports that Al-Adnani had suffered life-threatening wounds in a coalition air strike in January 2016 on the Iraqi town of Barwanah; rumor was that he had lost large amounts of blood in the attack.[290] The dispatches from the front lines were never confirmed. Abu Mohammed al-Adnani had simply vanished into thin air.

The hunt to locate and target Abu Mohammed al-Adnani was simultaneously carried out by multiple intelligence services. The espionage services—rivals working for different flags and possessing different capabilities—were determined to find that elusive needle in the haystack and take credit for killing one of the world's most wanted terrorists. The spies worked alone and in competition with one another—that was the nature of the intelligence business. It was a race against time to see whose source, which phone intercept, would reap the coveted reward of claiming credit for al-Adnani's death.

The GID, though, had an in.

It could take months, sometimes longer, for a GID case officer to identify and recruit an all-star asset. Patience wasn't a virtue in the spy game—it was a professional necessity. Thousands of man-hours, all sorts of risk and suitcases of cash could go into trying to locate and secure the services of an inside man—someone who could provide intimate, accurate and unimpeachable intelligence that no drone, satellite or phone tap could ever secure. The work was hit-or-miss; sometimes, the most cunning efforts ended in failure.

It wasn't easy being a case officer. Assets could be needy and emotionally draining. They came with all sorts of personal, professional and financial issues that made them susceptible to recruitment or sparked them to want to be an informant. Case officers were after intelligence, not friendships, and they were trained to do whatever it took, say whatever needed to be said and pay a fair price for the services of someone who could provide them with inside information. Still, the source's safety, his well-being and his family's security were of the utmost concern to the case officer. There was a lot of psychology involved in the relationship between handler and agent.

The case officer had to define the boundaries of the relationship with his source. He could never get too close, and was never permitted to reveal their real names or details about their personal lives. A case officer had to believe in his agent, though he didn't necessarily have to believe what he said. Assets often told their handlers what they wanted to hear rather than admitting that they had nothing to offer.

The case officer's mission was to squeeze information out of his source—cash, lies, temptation and threats were all resources that a successful handler could field. A man who will be called "Al" (a pseudonym) was one of the GID's most successful—and, some say, luckiest—case officers of them all.

All things considered, Al was a relatively youthful agent. He had joined the GID after 9/11, a benchmark in any intelligence service, and he was too new to even ponder retirement and life outside the organization. The stress of the job—alleviated only by caffeine and nicotine, was evident on his narrow face; his eyes always looked as if they could use an extra few hours of sleep, and a premature dusting of gray speckled his dark black hair and mustache. Al was soft-spoken and pleasant; he preferred to listen rather than speak.

Although Al, like most of his colleagues, was a former military man, he didn't sport commando-like biceps that some of the other agents boasted. He didn't have to. His strength was an acute understanding of the Arab mind-set and the intricate tradition-based thinking that dictated behavior. Al was a master of many of the regional dialects that were common in the countries that surrounded Jordan and could talk like a native with someone in Iraq as he could with someone in the Kingdom of Saudi Arabia. Al relished the mystique of the desert and knew the importance of clan and honor: he belonged to one of the larger Bedouin tribes in the kingdom.

Al's prized agent—in fact, one of the most important that the GID possessed in the war against the Islamic State—was codenamed "Suhayb." Suhayb ar-Rumi was a legendary figure in Islamic folklore. He was a slave in the Byzantine Empire who forged an important friendship with the Prophet Mohammed as he traveled throughout the Arabian Peninsula. The name has come to mean companion, or close associate, and was a fitting nom de guerre for Al's source. Suhayb was a one-time companion of Abu Mohammed al-Adnani.

Suhayb was one of the most important assets operating inside Syria against the Islamic State. All that security considerations will allow to be revealed about this amazing spy was that he knew al-Adnani in between his days as a firebrand religious student and his departure from prison to fight the Americans in Iraq. The GID has been careful not to disclose information about where Suhayb was from, what his role in the Syrian Civil War was and how he was able to provide intimate knowledge on al-Adnani's whereabouts to the GID. Details of his true name, age, appearance and residence are classified as top secret. No details were shared concerning how long it took to recruit Suhayb, or what the terms of his service were. No elaboration suitable for publication is permitted on exactly why Suhayb decided to do work on behalf of the GID. The only thing that one of the deputy commanders of the CTD would explain was that "Suhayb did the right thing."[291]

Suhayb was a known entity in the landscape of northwestern Syria. He could move about without suspicion in and around Aleppo, and because he was known as a companion of al-Adnani, Suhayb didn't attract the scrutiny of the Emni thugs who viewed virtually everyone as a threat. Because of his freedom of movement on a fluid battlefield, Suhayb was able to monitor Islamic State troop movements and deployments against Russian and

Iranian-backed forces and against the Syrian Kurds and the Free Syrian Army. But Suhayb's greatest contribution was his ability to recognize al-Adnani as he moved in the northern stretch of Syria disguised as a shepherd, a refugee or a woman. Satellites and drones could not pick out al-Adnani's eyes in a market crowd. Suhayb's reports were unprecedented in the up-close-and-personal details needed to isolate him and order an air strike to eliminate him.

Suhayb relayed to Al that al-Adnani's paranoia shifted following al-Shishani's death. Al-Adnani continued his masquerade to avoid detection, and he stayed in a different place every day, trying to move only under the cover of darkness. But, perhaps feeling threatened more than ever, al-Adnani remained in the part of Syria where he felt most comfortable—in the north near the Turkish frontier and near Aleppo.

Before the civil war Aleppo was Syria's largest city—larger even than the capital, Damascus. It was the third-largest metropolis in the region after Istanbul and Cairo. Once home to over four million inhabitants, only two million people remained in Aleppo by the summer of 2016. The city had been bombed into oblivion and resembled an apocalyptical nightmare of rubble and destruction. Antigovernment forces from a dozen or so different militias—including the Islamic State and the al-Nusra Front—controlled the city, though a siege by the Syrian government and Iranian-backed forces, together with Russian air support, had rendered Aleppo a wasteland. It was unsafe for the frontline forces, let alone an indispensable figure like al-Adnani. Binnish was only twenty-eight miles from Aleppo, but his Emni protection detail knew that his hometown was an obvious focal point of the Russian and coalition intelligence services, as well as Turkish spies and other traffickers in espionage. Al-Bab, though, a small town twenty-five miles to the northwest of Aleppo, was an ideal hiding spot for the Islamic State chief.

Sixty-thousand people lived in al-Bab, literally "the door"

in Arabic. The town had escaped much of the devastation witnessed in nearby Aleppo, even though it became an Islamic State
stronghold in 2013. Al-Bab was an essential stop on an ancient
trading line that connected Latakia on the Mediterranean with
Idlib, Aleppo, Manbij and Kobanî along the Turkish frontier.
Suhayb moved along the line with relative impunity and he was
able to take notice of the types of vehicles that al-Adnani's security detail used to ferry the Emni chief throughout the area.
The vehicles traveled in the direction of al-Bab and a specific
villa on the periphery of town. Al-Adnani had fallen prey to the
lackadaisical trappings of routine and overconfidence.

Al-Bab wasn't just a sentimental reminder of childhood inno-
cence to al-Adnani; it had become a critically important frontline command post in the Islamic State's attempts to push back
US-backed coalition attacks against key supply routes near the
Turkish frontier after the strategic town of Manbij fell and a
dedicated offensive to push out into the Euphrates Valley heart-
land commenced.[292]

In the wake of al-Shishani's death, the enigmatic al-Adnani
had become the face of the Caliphate—a man schooled personally by al-Baghdadi in Camp Bucca and who had personally
branded the Islamic State and then became its marketing genius.
But as coalition air strikes intensified al-Adnani preferred to
enjoy his status behind-the-scenes—especially after Paris when
he knew that the coalition intelligence services painted a bull's-
eye on his back. Al-Baghdadi didn't care. It was essential now
that al-Adnani show his face so that they could bolster the morale of the men on the front lines and reverse the fortunes of
the military effort.

Technology made the transfer of timely intelligence a relatively immediate undertaking—such as using text messaging
or emails—but these methods were not always considered safe.
The Emni bought off-the-shelf software to break into phones

and computers, and it employed some of the best hackers in the business—specialists who could unlock mobile phones and who could tap into what were believed to be impenetrable networks. Face-to-face encounters were always preferred.

The introduction of Jordanian personnel at coalition forward operating bases made such meetings somewhat easier. Tactical backup was built into such surroundings because Al or any other GID case officer could never absolutely tell if a source would suddenly betray them. As Allen Dulles, the first civilian director of the Central Intelligence Agency, explained, "Never take a person for granted. Very seldom judge a person to be above suspicion. Remember that we live by deceiving others. Others live by deceiving us."[293]

All exchanges between source and handler were fraught with danger and required extensive preparation. It didn't matter if the meet was an isolated sit-down, perhaps in the darkness of a desert night far from everything with the two men sitting over cups of Bedouin coffee brewing in a pot called a *bakraj*, or if it was held in a rented flat smack dab in the center of a city. The rendezvous always began with the case officer inquiring about the agent's health and his family's well-being, and a review of security precautions taken to make sure that he—or she—wasn't targeted by the Emni. It was standard practice that the case officer let the asset know that he was important but never essential; the human agent needed to know that he was one tile in a larger mosaic.[294] Most important, the face-to-face encounters allowed the case officer to expand on his questions. Information that might be trivial to Suhayb could be critical to Al in the assembly of an operational dossier of actionable intelligence.

The GID went to great lengths to verify Suhayb's information. Other sources were reviewed and examined; aerial reconnaissance imagery was double-checked. Suhayb's intelligence on al-Adnani made it all the way up the chain of command—to the head of the CTD, the deputy head of the service, Faisal

Shobaki and, ultimately, King Abdullah. The king relayed the heavily redacted information of al-Adnani's location to Jake and his deputy. All mention of Suhayb was removed from the report. The news, though, was encouraging—especially after Jake was informed that the source was unimpeachable.

Jake and his deputy rushed back from the meeting at al Husseiniya Palace to his office at Amman Station. Staffers were hurriedly assembled in the SCIF; encrypted messages were dispatched to Langley and to the Pentagon. Intelligence bonanzas always generated a lot of hammering on the keyboard; though they didn't know his name, Suhayb was referred to as a "reliable source."

The electronic back-and-forth ultimately reached the coalition air force's Combined Air Operations Center at al-Udeid Air Base in Qatar. The targeting process took time—attack aircraft had to be assigned, refueling aircraft designated, and CSAR elements prepared and readied to respond should one of the pilots be forced to eject like Moaz less than two years earlier. Coalition protocols were changed following Moaz's capture. Combat rescue assets were no longer prepositioned a four-hour flight away. They were nearby and ready to deploy. It was essential to make sure that the Islamic State didn't get its hands on another coalition pilot.

The sortie was scheduled for August 30. The international campaign against the Islamic State was completing its second year. The mission to take out al-Adnani, it is believed,* was assigned to the US Navy.

The Nimitz-class nuclear-powered aircraft carrier USS *Dwight D. Eisenhower* (CVN 69) was the flagship of a US Navy carrier strike group that included one cruiser, ten destroyers and frig-

---

* "Officially" speaking, CJTF-OIR does not report the number or type of aircraft employed in a strike, the number of munitions dropped in each strike, or the number of individual munition impact points against a target.

ates, and an air wing consisting of close to seventy combat air-craft. A carrier strike group was the largest single operational unit in the United States military's order of battle and defined how Washington, DC, projected power around the world. The *Eisenhower*, nicknamed "the Mighty Ike," was on duty in the Persian Gulf to support the Inherent Resolve coalition when the al-Adnani mission was assigned to the F/A-18Cs of the Strike Fighter Squadron 131. The pilot briefings were at dawn. Ord-nancemen in long-sleeve red tunics fitted the aircraft with the bombs and missiles that would be used in the attack. Hook run-ners in green shirts, the busiest crewmen on the deck of a carrier, maintaining the catapults and arresting gear, helped launch an E-2C Hawkeye AWACS (airborne warning and control system) aircraft to clear the flight path over Iraq and then into Syria. The Hornets were airborne a half hour later. The flight to Syria, in-cluding refueling, took over four hours.

Abu Mohammed al-Adnani arrived at the house in al-Bab before dawn. His security entourage had prepped the location with guards throughout the perimeter and in the surround-ing streets and roadways; heavily armed men were positioned to protect against any attempt by special operations units from launching a raid to kidnap or kill the Islamic State deputy chief. The summer sun and the prying eyes of airborne reconnaissance made it prudent to remain indoors during daylight hours. The house was a hillside villa of considerable size and comfort, opu-lent enough for a man of his stature.

Specific details of the air strike remain classified. It is known, though, that the 500-pound bombs hit the villa at midday, caus-ing the structure to erupt into a massive fireball that was fol-lowed by a mushroom cloud of black smoke and debris. Nothing and no one could have survived the blast; a huge crater was cre-ated where a multistory villa once stood. It was a fitting end to the man who ordered the immolation of First Lieutenant Moaz al-Kasasbeh twenty months earlier.

The Hornets returned to the deck of the *Eisenhower* before sunset. The mission had been precise and executed perfectly.

The Islamic State broke the news of al-Adnani's death within hours of the bombing. The F/A-18 strike that killed him also killed several key regional commanders whom he had been meeting with, and the media center in Raqqa found it impossible—and perhaps in their best interest—to publicize word of the latest high-profile martyrs. The Islamic State, through its friends in the Amaq News Agency, released a statement, as translated by the Site Intelligence Group: "After a journey filled with sacrifice and defense against disbelief and its party, the lion-like knight Abu Muhammad al-Adnani al-Shami [the Syrian] dismounted, to join the caravan of martyred leaders, the caravan of the heroes who waged jihad and were patient on the command of Allah, and preserved against the enemies of Allah, and remained garrisoned on the fronts of Islam, and spoke the truth aloud while the death lay in wait for them."[295] The Islamic State did not fabricate word of its top-level commanders killed by the coalition.

Abu Mohammed al-Adnani was irreplaceable. Al-Baghdadi knew it. The new guard of leadership did not have the common bond of fighting American and Iraqi forces as part of the Zarqawi network; the new vanguard did not share the harsh jihadist academy years of Camp Bucca and Abu Ghraib. "There was a difference between the old guard who built an organization and marched it across the Middle East," a retired GID officer who chose to remain anonymous explained, "and the up-and-comers from all over the world who arrived after the fact in order to share in the spoils."[296] Al-Adnani was the voice of the movement. His ability to motivate and horrify was his alone, and it was an intangible that the Islamic State would find hard to replace at a time when it needed an injection of zeal the most.

There were attempts by many involved in the war against the Islamic State to spin the narrative surrounding the bombing strike for their own benefit. As al-Bab was only sixteen miles from Dabiq, the legendary town in Syria where the final battle between Islam and the nonbelievers is to be fought, individuals with close ties to the Islamic State hinted that al-Adnani had been killed there while inspecting his troops in the field and resisting non-Muslim forces.[297] There were rumors published by the *New York Times* claiming that al-Adnani had been killed by the CIA and Special Forces in a drone strike on the vehicle he was riding in.[298]

The Turks took credit for the air raid that killed al-Adnani, as did the Assad government.[299] The Russian ministry of defense issued a detailed statement claiming that al-Adnani had been killed in the bombing of the Aleppo area by Sukhoi Su-34 twin-seat all-weather fighter-bombers; there were some Russian claims that al-Adnani had been taken out by a sniper shot.[300]

In Paris, the DGSI released al-Adnani's death notice before the story was broken by France's vibrant press. For the French, even the mere hint that its military and intelligence services had played a role—even if minor—in the obliteration of the man who masterminded the Paris attacks was of vital importance to promote their role in the war on terror. The DGSI, in getting ahead of the story, was preparing the public. There were bound to be revenge attacks, and since al-Adnani had such a strong network operating in Europe, France again was a likely target.

The Pentagon claimed credit for the strike though it was wary of making another al-Shishani type of premature error. "Today coalition forces conducted a precision strike near al-Bab, Syria, targeting Abu Muhammad Al-Adnani, one of ISIL's most senior leaders," said spokesman Peter Cook. "We are still assessing the results of the strike, but Al-Adnani's removal from the battle-field would mark another significant blow to ISIL."[301]

The RJAF liaison officer assigned to al-Udeid Air Base in Qatar was the first Jordanian to learn that the mission in al-Bab had been a success. He sent his reports through channels, up the chain of command, to the respective units and desks that handled such information. News quickly spread to the armed forces—by the time the Jordanian Armed Forces were sitting down to dinner on the night of August 30, the pundits on Al Jazeera were already talking about al-Adnani's death.

Brigadier General Ghassan,* the Muwaffaq Salti Air Base commander, and Lieutenant Colonel Ali, the No. 1 Squadron commander, watched the news that night with reserved delight. Moaz al-Kasasbeh's commanders had no idea about the GID role in sealing al-Adnani's fate, though they quietly suspected—without knowing anything more than a wink and a nod—that Jordan's spies were involved some way and somehow along the chain of events that led to the targeting of the man who executed their brother-in-arms. The news brought a sense of relief and closure to the men at the base—especially to the F-16 pilots who were preparing for their bombing sortie the next morning. Because al-Adnani had turned Moaz's capture into such blood-chilling cruelty, there was a sense that a measure of vengeance had been achieved. "It was an enormous weight taken off our shoulders," one of the pilots commented. "For us here in the squadron, in particular, it was something like closure. But I sure as hell hope the bastard suffered."[302]

The officers assigned to the Moaz Task Force wondered the same thing. They had to watch the film of Moaz being burned alive over and over again as they investigated the crime and searched for those who perpetrated it. Moaz's suffering in the cage was a bitter reminder to the men who hunted al-Adnani that the epitome of evil had masterminded an act of cruelty that resulted in the unspeakable suffering of one of their own. They wondered how long it took the bastard to die. They all hoped that his end was incredibly painful and humiliating and that it took him a long and excruciating time to leave this world.[303]

---

* A pseudonym.

The elimination of Abu Mohammed al-Adnani was an epic victory for the GID. The mood at headquarters reflected that achievement. The men celebrated, of course, and boxes of sweets were shared in a conference room amid smiles and a sense of incredible accomplishment. Details of exactly how the intelligence on al-Adnani was gathered and then processed remain top secret—the very existence of Suhayb, let alone his true identity, was known only to a handful of men. Al and his team would receive commendations in their service jackets, but the full story of the incredible work that a handful of men did under incredibly dangerous conditions would have to remain hush-hush. Such was the nature of the spy game.

King Abdullah received news of al-Adnani's removal in person—GID director Shobaki, CTD boss Abu Haytham and a team of CTD commanders briefed the monarch on the operation at his military office. The king was overjoyed by the news; vengeance wasn't just pleasing on a visceral level, but strategically it was a turning point. "Killing al-Adnani was removing the head of the snake," the Jordanian monarch would later comment.[304]

The summer of 2016 had been a productive one for the GID—first al-Shishani and then al-Adnani. Abu Haytham allowed his intelligence officers to relish in the thrill of victory. But he reminded his men that there were still three very guilty men lurking about Syria and Iraq and that the efforts of the Moaz Task Force would continue until all three were dead. The CTD boss also reiterated the fact that the GID was on the front line in the overall war against the Islamic State, and unless the intelligence was sharp, accurate and reliable, the coalition would find it hard to stay one step ahead of the khawarij and put an end to the war and the ideology the jihadists brought with them.

The photos of al-Adnani were gathered and placed inside a folder that would be locked away for safekeeping. One photo of al-Adnani was removed from the folder, as was a wanted poster featuring al-Shishani. The photos of the two Islamic State leaders who had been killed were taped to the whiteboard and a red *X* drawn across their faces. There were scant images of the three remaining men. If a photo existed, it, too, was taped underneath the marked photos of their dead bosses. There was no time to waste—the war was entering its third year.

Days after al-Adnani was killed, the Moaz Task Force met to plan and coordinate a strategy to end the lives of Omar Mahdi Zeidan, Abdul-Hai al-Muhajir and Abu Khattab al-Rawi. Forty days later the offensive to recapture Mosul from the Islamic State began. Al-Adnani's death had been a game changer.

19

Dirty Battles, a Dirty War

The two most powerful warriors are patience and time.

—Leo Tolstoy

IN JUNE 2014, Sirwan Barzani, a general in the Iraqi Kurdish Peshmerga militia, drove his Toyota Land Cruiser to a hilltop near his headquarters close to the town of Makhmur and watched what was left of the Iraqi Army flee for its life. The Islamic State was now marching straight for the heart of Iraq's Kurdish Regional Government, promising to carry out massacres against the Kurdish and Christian population that would eclipse the horrific bloodletting perpetrated by Saddam Hussein's army years earlier. The jihadists closed in on the city of Erbil, the capital of the Kurdistan Regional Government (KRG), an autonomous area that was as close to an independent state as the Kurdish people had ever had. After all, as Henri J. Barkey, professor of international relations at Lehigh University, explained, "The Kurds are the largest ethnic group in the world who do not have a state of their own."[305] Flush with oil and foreign investment, and energized by a welcoming Kurdish disposition, the KRG had become a shining success. Now, ISIS threatened it.

The word *Peshmerga* literally translates into "those who face death," and the forty-four-year-old Barzani, a member of the extended clan that has dominated the political and military landscape of Iraqi Kurds for generations, has been at war since he was old enough to carry an AK-47. When Barzani was thirteen years old, Saddam Hussein's henchmen hanged his father; he had claimed that the Iraqis killed eight thousand members of Barzani's extended clan.[306] The Iraqi Kurdish landscape was littered with stories of massacres and entire families destroyed. In one of the Iraqi Baath Party's many campaigns against the Kurdish people, the world was horrified when, in 1988, Saddam Hussein's air force launched a chemical attack against the town of Halabja, gassing everyone in town, and killing close to five thousand people in a nightlong blitzkrieg; thousands more were seriously wounded.[307]

Sirwan Barzani commanded what was known on the military map as Sector 6, an area stretching from Mosul south toward the oil fields of Kirkuk. To the outside eye, Barzani's contingent of Peshmerga didn't look high-speed and low drag. They were known as the "Black Cats," and they consisted of men and women who looked as if they had been outfitted from boxes of discarded uniforms and secondhand load-bearing gear. But appearances didn't win battles. The Kurds were formidable combatants when it came to fighting with their backs to the wall. They were legendary mountain fighters. A popular proverb among the Kurds was that they had no friends but the mountains. The Peshmerga offered to help the Iraqi central government hold Mosul, but Baghdad refused. The Shiite commanders thought that they could hold the city on their own. The Iraqis were arrogant, and in viewing Iraq through a sectarian prism, they had squandered control of the country's second-largest city.

On that hilltop, swept by the unbearable hot winds of the oncoming summer, General Barzani watched as a complete army, several divisions full of men and equipment fled on the roadways leading toward Erbil and Baghdad. The Iraqi Army, according to the 2005 postwar Iraqi constitution, wasn't supposed to be in Kurdistan at all,[308] but these men—and their commanders—were running for their lives without ever having fired a shot, and the constitution didn't mean a thing when Islamic State forces were in hot pursuit.

Barzani had 150 fighters and eight US Navy SEALs assigned to his command by both USSOCOM and CENTCOM to hold the line. "Da'esh knew that it could terrorize a city," he recalled, "but to take over this area they'd have to get past the mountains and us."[309] Barzani and his men held the line, though. They followed classic military protocol and built hastily dug berms out of the hard earth. Islamic State forces advanced on the highway, riding atop captured American-made M1 tanks, MRAPs, HMMWVs and the Toyota catalog of all-terrain vehicles and pickups. By building barriers that were two meters high and three meters deep, Barzani and his Peshmerga forces hoped to slow down the Islamic State advance and bottleneck them into preordained kill zones.

"We didn't have sophisticated antitank weapons or anything that resembled artillery," a Peshmerga officer in Sector 6 explained, "so we relied on the weapons that we had and what slowly came to us from the Americans and, later, from the coalition members. Eventually, Norwegian, Dutch, and Australian special operations support came to help us."[310] Operators from CANSOFCOM, the Canadian Special Operations Force Command, including those from the tip-of-the-spear Joint Task Force 2 counterterrorist and hostage-rescue unit, were rushed to Kurdistan to assist the Peshmerga—especially in small-unit tactics and how to call in close air support and air strikes.[311]

The Islamic State's most effective weapon was vehicle-borne improvised explosive devices that rammed through checkpoints and barriers in suicide attacks. When German special forces arrived and brought the MILAN antitank guided weapons with them, the Peshmerga were able to engage the suicide truck bombs at ranges of close to two thousand meters away. "Islamic State engineers rigged their vehicle bombs with armor plating, but it didn't matter," General Barzani recalled. "The MILANs sliced through everything. Coalition aircraft soon began to provide air cover for us, attacking Da'esh columns as they moved toward our lines. It was the first time in Kurdish history that combat aircraft in the sky were helping us and not attacking us."[312]

Facts on the ground had changed dramatically in the two years since the Islamic State pushed on Makhmur and Erbil, after Abu Bakr al-Baghdadi entered the Great Mosque in Mosul to declare the Caliphate and bring fear and foreboding to the capitals of the Middle East. The Islamic State was now in full retreat.

By the fall of 2016, the Caliphate had shrunk on all fronts. It was harder for the Islamic State to fund its attempt to create a nation stretching across oceans and time zones. Vast stretches of territory were lost in battle—cities, villages and roadways in Syria that had black banners hanging from town halls and from power lines were now liberated by Assad's army and a coalition of Kurdish and other factions. Oil fields and oil routes were lost.

In Iraq the army had scored considerable successes against Islamic State gains in Anbar Province and elsewhere. Coalition special operations units helped to lead the way; more conventional formations from the US Marine Corps and the 101st Airborne Division came to the rescue, as well, providing command-and-control guidance and artillery support. On the ground, operators in their durable digital-pattern camouflage fatigues were everywhere. Coalition combat aircraft dominated the skies.

With al-Shishani and al-Adnani rendered to the long list of Islamic State commanders no longer able to muster morale and military strategy, the time to retake Mosul was at hand. Liberating Mosul was seen as an essential move to end the conflict once and for all.

The Iraqi Army no longer thought it wise to dismiss the Peshmerga. Nor did Baghdad believe that its armed forces were capable of defeating the Islamic State all on their own. Kurdish forces now constituted the front line, face-to-face with the

Islamic State, on the outskirts of Mosul; American and coalition forces, primarily special operations units speaking a Ro-
setta stone of languages, provided tactical and logistical support
to the Iraqi Army and police units mustering outside the city.

The Peshmerga assembled a force of forty thousand fighters
who waged a difficult and costly advance into the villages and
farming communities around Mosul, breaking through Islamic
State defenses and seizing scores of prisoners. Islamic State forces
used their most effective weapon, suicide car bombs, against the
"Pesh," as US Special Forces called the Kurdish militia, resulting
in scores of dead and wounded on both sides. The Peshmerga
cleared the main road between Erbil and Mosul and neutralized a
staging area roughly two hundred square kilometers wide, from
where the final push into the city would be launched.[313] Islamic
State fighters booby-trapped every building, every corner and
every intersection. "IEDs were everywhere," Lieutenant General Jamal Mohammed, the Peshmerga chief of staff, explained.
"Small Islamic State units used ingeniously designed devices to
kill and wound scores of Kurdish fighters. The Islamic State
used drones and other homemade explosive devices. They had
an entire industry dedicated to devious tools of destruction."[314]

Kurdish fighters provided an opening to the Iraqi military to
position itself for the final assault. The cost in dead and wounded,
though, was staggering.* The Islamic State was going to make
the Kurds and the Iraqi Army pay for every inch of ground they
took back. The fighting was nightmarishly close-quarter; Islamic State forces set oil drums alight that would produce thick
clouds of acrid black smoke to obscure the view from above for
the satellites, the drones and the fighter aircraft flying in support of the multinational and multimilitia coalition campaign.

But by the end of October, the Peshmerga were in control of the
western areas of the city.[315] Forces swept around from Sinjar, lib-

---

* According to Lieutenant General Mohammed, the Peshmerga lost 1,780 killed
in action, and 11,000 wounded.

erated a year earlier, and then moved to the boundaries of Mosul, effectively cutting off the city from the Islamic State's spiritual capital of Raqqa. The Islamic State, and the wars to defeat it, have redrawn maps and on-the-ground realities in the Middle East, erasing boundaries and ethnic hierarchies that existed since the end of the Ottoman Empire and even centuries before. For the first time in their history, the Kurds found themselves in control of large populations of Sunni Arabs and other minorities.

The Kurdish and Iraqi-led military move on Mosul created a new flow of refugees and a new humanitarian crisis. The Islamic State had instituted a phone-book-size list of laws while in control of the city, including the legal enslavement and sexual imprisonment of all non-Muslims; Sunnis now in Kurdish care feared that similar laws would be instituted against them in revenge. Instead, aid came flowing from around the Middle East. Food, tents and medical supplies poured in from Saudi Arabia, the Emirates and from the Kurdistan Regional Government.[316] NGOs established pipelines of supplies to feed and care for the new wave of displaced persons to emerge from the latest round of bloody combat. The humanitarian ramifications of the back-and-forth of fanatic conquest and scorched-earth liberation produced a never-healing wound of fear and misery.

At one of the strategy meetings prior to the commencement of the offensive, Barzani met with his Iraqi and coalition counterparts and offered his input. "Leave the Da'esh forces an escape, an opening, through which they can escape," he urged the generals. "They aren't looking for a fight to the end, they want to try and flee with their lives and their money. And it's when they escape that they can be finished off once and for all."[317] General Barzani understood that, if trapped, the Islamic State would fight to the death. Islamic State fighters had committed unspeakable crimes, and the locals—the Kurds, the Christians, the Yazidis and the Shiites—would want their vengeance. "Give them an out," the Peshmerga commander offered, "and when they escape, coalition aircraft will incinerate them like they had done to Saddam Hussein's army fleeing Kuwait at the end of the First Gulf War in 1991."[318]

The Iraqi generals scoffed at the idea. They wanted their magnificent victory. Inherent Resolve coalition estimates were that there were still one and a half million civilians in Mosul who were being held hostage by the Islamic State and used as human shields. The Sunni population, fearing that the Shiite-dominated Iraqi military would take reprisals against them once the fighting ultimately ended, braced for the worst-case scenario. The problems that helped create the Islamic State hadn't disappeared in the two years of the Caliphate. Old hatreds and fears didn't disappear, even under the yoke of the black banner.

The offensive to retake Mosul began on November 1, 2016, when Iraqi special forces, assisted by US, British, French and Canadian special ops teams, entered the city. The Iraqi troops, especially those from the elite Golden Division, went into the city armed with inkjet color printouts of wanted men, a roster of those who had massacred and raped, looking to arrest suspects and make the guilty atone for their sins.

It had taken less than ninety-six hours for the Islamic State to conquer Mosul. Even the most bombastic Iraqi commanders knew that it would take weeks—perhaps months and even years—for the black flags to come down once and for all.

At the US Embassy in Baghdad, at the CIA station, and at Amman Station and GID Headquarters, the intelligence officers worked off of a different list of wanted men. The Americans were after the local and regional commanders, who knew where the bodies were buried and where the money was hidden. Mostly they wanted to help eradicate the Islamic State recruiters and the financiers, the bomb builders and the overseas planners, to ensure that the Caliphate would disappear into the ashes of history. The GID was also concerned about the future of its two neighbors to the north and wanted to make sure that

Abu Bakr al-Baghdadi's contagion was quarantined, destroyed, and that it would forever lack the energy to rise again.

Five days after the offensive began, the American people went to the polls to elect a new president. Donald J. Trump, who ran on an "America First" nationalist platform, was seen as someone who supported Russia's position in the Middle East—especially in Syria—and his surprise election created more questions and uncertainty for the monarchs, emirs, prime ministers and presidents who ruled the Middle East. Uncertainty, all too often, was a prerequisite for more bloodshed.

Meanwhile, King Abdullah paid close attention to the events in Mosul. So, too, did his generals and his spymasters. There were three Islamic State leaders whose names still appeared on the GID's ledger. Omar Mahdi Ahmad Zeidan, perhaps the most dangerous of the three to Jordanian security and certainly one of the most despicable, was held up in the western portion of the city.

Omar Mahdi Ahmad Zeidan, known in Islamic State circles by the nom de guerre Abu Monzer al-Urdoni, "the Jordanian," was a passionate advocate with powerful oratory skills. His dynamic on-air presence put him on television when the world was at war with bin Laden, and he became one of al-Qaeda's top media advocates and key spiritual recruiters, justifying heinous acts of murder with eloquently crafted, bizarre interpretations of the Koran and religious edicts. AQAP wanted religious indemnification for the violence, and his delivery was more important than the content. Zeidan used his on-air preacher skills to convince al-Baghdadi to make him the *diwan*, or head, of the Islamic State's education division.

Zeidan was responsible for the education of those who swore allegiance to al-Baghdadi. Many of the combatants were new to the fundamentalist path of the religion as interpreted by the Islamic State and al-Baghdadi's divine council. A good number of the foreign fighters, especially those from Europe, might have come from

Muslim homes but they had spent more time on street corners selling or using dope than they did praying inside a mosque; most didn't speak Arabic of any coherency. Sculpting the minds—and refining the faith—of this frontline Caliphate army fell on Zeidan's shoulders. He was masterful at it. He convinced the followers that rape, enslavement, theft and murder for the sake of conquest were all part of the Prophet's plan. Zeidan was as responsible as any member of the ruling council for the medieval-like barbarity that the Islamic State exhibited in the region and beyond.

But more than anything else Zeidan hated the country of his birth. He wanted to spark a revolution inside Jordan and turn the kingdom inside out. When Moaz al-Kasasbeh was pulled out of the water that cold winter morning, the airman provided Zeidan with an opportunity. By setting an example of the pilot, the Islamic State could manipulate sympathies it already had in the kingdom and, at the same time, achieve something that had not been done in ages—unite the Sunni Arabs into a common cause that would set the Middle East on fire. Moaz, Zeidan urged members of al-Baghdadi's ruling council, represented the worst of American and Western efforts to corrupt the Arab world. The corrupt nations that accepted peace with Israel and formed alliances with the United States had caused a schism inside the Sunni world. Making an example of Moaz, Zeidan argued, would unite the Sunni Arabs and expand the reach of the Caliphate across Jordan, Israel and the Arabian Peninsula.

When al-Adnani suggested burning Kasasbeh as an unforgettable gesture of unwavering strength, Zeidan was responsible for providing the religious legal argument that justified the fiery execution, even though burning people alive was forbidden in Islam.

And because he was so calculating in manipulating the tenets of Islam and desecrating its laws for the Caliphate's own perverted purposes that many inside the GID found Zeidan to be a far more despicable target than al-Adnani. They wanted him dead. Zeidan, though, struck first.

On December 18, 2016, police and gendarmerie forces in the southern Jordanian town of Kerak responded to reports of an explosion in a building near the entrance to Kerak Castle, one of Jordan's top tourist sights and the largest Crusader-era fortification still intact in the Middle East. It was Christmastime, when tourists traditionally flocked to Jordan, Israel and the Palestinian Authority to celebrate Christendom's presence in the towns and temples where Christ walked. Christmas was always wet and cold in Kerak, and a bone-numbing rain came down as policemen pulled up to the reported location where people said they heard a blast. Suddenly four men in their twenties and thirties emerged from a nearby building armed with satchels and AK-47 assault rifles. They pinned down the police and then ran into the castle, firing indiscriminately as they ran.

The four gunmen were native sons of Kerak and members of a Zeidan-inspired cell that had been preparing for a much larger terrorist strike to coincide with the Christmas holiday. They had intimate knowledge of the layout of the castle. A raging gun battle erupted in which tourists were caught in the crossfire; a Canadian national was killed. Operators from CTB-71, the Jordanian military's counterterrorist and hostage-rescue unit, were called out.

The operation to liberate Kerak Castle lasted hours. Thousands of rounds were exchanged in a ferocious battle that lingered throughout the sprawling Crusader-era fortress. Ten people were killed in the battle, including seven Jordanian soldiers and policemen. All of the terrorists were killed, as well; another was fatally wounded in an arrest operation days later when the GID wrapped up members of the nascent terror cell.

The cell had taken the country—and the GID—by surprise. Sentiment in the area had seemed vehemently anti-Islamic State— the governate was where Moaz Kasasbeh was from. The incident at Kerak was a sober reminder to the GID—and to the rest of the Middle East—that the Islamic State would continue to strike outward while all eyes were aimed on Mosul. Still, as one former member of Jordan's intelligence community stated, "the Da'esh cell was relatively amateur, and did not carry out a catastrophic attack, nor did it change public sentiment one bit."[319] But there were worries in Washington, DC, and in Amman Station that perhaps, at this late stage in the war against the Islamic State, al-Baghdadi would try one last-ditch gambit to strike at the kingdom. And what happened in Kerak was of great concern to King Abdullah.[320]

Islamic State forces in Mosul had managed to hold for five months by the time the winter rains came to an end with the sparkle of sunshine that introduced the warmth of spring to the city. What General Barzani had feared most ultimately transpired—Mosul became an urban warfare nightmare for the Iraqi Army. The close-quarter fighting along with the close-quarter coalition air support turned the city into a labyrinth of rubble, shell craters and twisted rebar. Houses, then streets and ultimately neighborhoods were turned into Mad Max–like scenes of devastation, and that was good news for the Islamic State—it was always easier to defend military positions when camouflaged by rubble. The debris was ideal for the concealment of IEDs; Islamic State snipers, particularly the Chechens who were known for their sharpshooting acumen, found an endless array of hides where they could pick off Iraqi soldiers from five hundred meters out and stall a whole brigade's advance. The winter rains had turned the streets slippery with mud and human waste; shell craters became pools of muck. And somehow, in the death and destruction, families that were trapped inside the kill zone had attempted to find food and secure shelter. Many didn't. Bodies were buried under multiple layers of rubble, left to decay amid the endless firefights throughout the city.

The civilian population of Mosul had become a million-strong human shield protecting approximately six thousand Islamic State gunmen. Most of the Islamic State's top leadership had abandoned Mosul the moment Peshmerga forces pushed in from the west.[321] But Omar Mahdi Ahmad Zeidan—and his phalanx of Emni security personnel—remained behind to fight and to rally the forces.

The Emni's effectiveness had come under question following the assassinations of al-Shishani and al-Adnani. Emni commanders couldn't comprehend how coalition airpower managed to get both al-Shishani and al-Adnani in the span of six weeks. Both men had adhered to the most stringent tradecraft requirements, staying offline, never using their phones. Both of the men had been paranoid in who they met with and who knew of their future movements. Yet they had been found.

With al-Adnani dead, a power struggle emerged in the ranks of the Emni. Men who wanted their chance to be at the top played a dangerous and desperate game of politics. Al-Adnani had been an agent of cohesion inside the Emni. He ruled with such an iron hand, trusting no one and suspecting everyone, that his lieutenants and rivals were always too off-balance to try to stage a coup of some kind. Now that he was dead, an orgy of suspicion consumed the counterintelligence divisions. "Purges were part and parcel with illegitimate terror organizations as fearful of its enemies on the outside as they were from its enemies from within," a retired GID officer who worked the Kurdistan front explained, "and the Islamic State was an entity that feigned being built on faith but was really constructed on a foundation of pure fear."[322]

Coalition intelligence chiefs based in Amman, Ankara, Erbil and Baghdad watched happily as the Emni began to unravel along sectarian and religious lines. The Syrians inside the Emni suspected those from the Caucasus of having been infiltrated by Russia's GRU and FSB; the Georgians, the Tajiks and the Chechens inside the organization's ranks suspected the North Africans of possible ties to the Egyptian intelligence service, the CIA or the countless other spy services that ran a brisk business inside the Caliphate; Iraqis were suspected of supplying the Iranians with information, and anyone who suddenly flashed cash was believed to be on the CIA's payroll.

Emni investigators were stymied, though, because they could find no common human link between the two men—no singular aide or minister, other than those on the ruling council. The Emni, therefore, did what all fanatical organizations do well: they lined innocent men and women up against the wall. Hundreds were executed in the rear of Mosul's police stations where the Islamic State ran day-to-day life. Those who confessed their sins—real or imagined—were strung up by the neck or other extremities from light poles. The Emni executions did nothing to root out traitors in the ranks. Those who had supplied the GID with the vital intelligence used to kill al-Shishani and al-Adnani were long gone. The executions were also terrible for morale, especially as the fighting took a devastating toll on Mosul's Islamic jihad.

But while the Emni looked inside Mosul for traitors, Zeidan was betrayed from far outside the battle lines.

Once it appeared that the coalition war against the Islamic State would be successful, individuals with ties to key Caliphate officials were easier to recruit and press for information. "People who aren't fanatics are less likely to risk everything for a losing and fanatical cause," explained Salah,* a GID officer assigned to the CTD, "and someone close to the Zeidan inner circle thought it wise to assist us before any and all offers were off the table. One of the unique attributes of this organization is that we can reason with people, talk to them as Arabs, as Muslims, as people who share the same common background and aspirations for the same type of future, that other services have to use money or coercion for."[323]

The source, as he *or* she needs to be called,† was someone who was close to Zeidan, though it cannot be said if the person was

---

* A pseudonym.

† Due to security concerns, the name, age, sex and relationship to the target cannot be revealed. All that can be said is that the source was someone with a close-knit sort of acquaintance to Zeidan.

friend, family or former student. It didn't matter. Once Zeidan and the source established communications, the GID case officer took charge of the operation.

The source spoke with Zeidan through a coordinated system of communication that included staying off existing GSM networks[324] and using a "cutout," or mutually-trusted intermediary; such practices, including making sure that all calls and internet traffic were done via satellite dish and not phone companies, were vital to securing anonymity and staying one step ahead of the aircraft, the suicide drones and the teams of the top-tier special operations units that capitalized on such tradecraft errors.

Once the link was established, the GID monitored every communication between Zeidan and the source. The intelligence officers who worked the operation relied on a team of in-service psychologists to prepare text messages that served their purpose that were wordsmithed with the cadence and syntax that wouldn't arouse suspicion; they even scripted the words needed in phone conversations. Zeidan, having been schooled by jihadist cells in both Yemen and Syria, was well versed in suspicion. The back-and-forth, the questions and the responses they hoped to elicit, had to appear to be innocuous and based out of reverence and concern. The objective—always—was to get the target comfortable. Once the rapport was established and the times of the calls and SMS messages set, it was easy to paint the electronic bull's-eye on Zeidan's head.

The CIA liaison officers arrived at GID headquarters one very raw morning at the end of February 2017. Amman was cold in the winter; snowstorms that turned the Jordanian capital into a slippery obstacle course of stone and ice were not rare. The liaison officers removed their coats and sat down for the obligatory Bedouin coffee followed by tea and sweets. The protocols of the get-togethers were nonnegotiable. A sealed file on Zeidan was handed to the Amman Station liaison team as they ate cookies. There were pages in the dossier that were heavily redacted. The

GID officer chairing the meeting explained, as circuitously as he could, why the pages consisted of so many black lines across key lines of text. The CIA officers understood.

The information was transmitted back to Langley and, most likely, to Baghdad Station and the Agency's representative in Erbil. Both the CIA and the GID maintained close ties to the Peshmerga's intelligence service, the Parastin u Zanyari, and the standing policy was to keep the Kurds in the loop. The Kurds had developed an intricate network of spies inside the Islamic State, as well—efforts enhanced by the intelligence-links with the co-alition partners. "We tried to know what went on with our en-emies," a Parastin u Zanyari officer explained while sitting over a wall-size map in Erbil. "For an underdog, knowing was as im-portant as fighting, and you couldn't fight without knowing."[325]

The Kurds played a vital role in directing coalition aircraft to their targets over Mosul. The skies above the city were con-stantly full of aircraft streaking in from high altitudes to drop precision ordnance on Islamic State targets all over the city. The roar of supersonic jet engines competed with the crackle of gunfire and the deafening blasts of suicide bombs. The aircraft sported a dozen different markings, representing the air forces and navies of many nations: America, Canada, Britain, Austra-lia and France, to name a few.

On March 7, 2017,* one of those sorties utilizing the target-ing data made available by the source's intelligence dropped their ordnance on a house in the old city of Mosul where Omar Mahdi Ahmad Zeidan and his Emni detail were holed up. The guided ordnance obliterated the concrete skeleton of the build-

---

* The GID provided the date of March 7, 2017, for the strike that killed Zeidan. Coalition press releases, and statements by commanders, listed air operations that day over Mosul as consisting of "five strikes engaged five ISIS tactical units; destroyed nine fighting positions, five VBIEDs, four vehicles, three roadblocks, two mortar systems, two tactical vehicles, two rocket-propelled grenade systems, a supply cache, an artillery system, a sniper position, a UAV launch site, and a VBIED factory; damaged 23 vehicles, 10 supply routes, and a tunnel; and suppressed two mortar teams."

ing and burrowed a ten-meter-deep hole of devastation. The crest of the black mushroom cloud could be seen for miles outside the city. There were no survivors in the apartment block where Zeidan had set up his ministry and command bunker.

It was a fitting end for Omar Mahdi Ahmad Zeidan. He had looked at the terrified face of a beaten young pilot and saw a convenient opportunity to set Jordan and the region ablaze by sanctioning barbarism as a measure of faith. His end was atop the bonfire of misery and suffering that those who followed in Moaz's steps had pledged to snuff out.

The fighting continued in Mosul. It had taken the Iraqi Army—supported by a multinational First World force of the most powerful armies on the planet—five months to inch its way into a small sliver of the city. Islamic State fighters, in small bands of suicide squads, were determined to remind their Iraqi attackers that they had seized the city in a matter of days, but retaking Mosul would take an eternity and cost the Iraqis an endless graveyard of dead soldiers. A colonel named "Steel" for his tough-as-nails approach to the fighting articulated the carnage best. A section commander in the Iraqi Army special operations force known as the Golden Division, which was fighting street by street and house by house in eastern Mosul in the notorious Gokjeli section of the city, Steel defined the fighting thusly: "An army can fight against a regular army in the usual way, tank against tank and soldier against soldier, but what is happening here is a dirty battle against dirty people."[326]

20

Dawn before Daylight

Perfect courage is to do without witnesses what one would be capable of doing with the world looking on.

—François, Duc de La Rochefoucauld

ON MARCH 31, 2017, King Abdullah issued a royal decree naming Major General Adnan Issam al-Jundi as the new GID director.[327] The mild-mannered al-Jundi had left the service a year earlier but came back to the GID to help reestablish order and security for Jordan the day after there was no Islamic State. Al-Jundi had spent much of his professional career inside the ranks of the GID—from rookie intelligence officer to senior command—and had seen the service grow and develop a greater regional and international importance as the global war against terrorism intensified.

Al-Jundi was known to be fiercely loyal to the king and to the organization, but most important, he was a man of unimpeachable honor whose word could be banked on in any currency. Such reputations were critical in the top-level relationships that the GID enjoyed with other allied—and sometimes not-so-friendly—services.

Al-Jundi's appointment was seen more as a strategic rather than a tactical move. As the noose around Mosul tightened and casualties—both civilian and military—soared, a relieved Middle East emerging from the horrors of the Islamic State appeared at hand. Iraq's fractious political system would have to be repaired; eventually, the fighting in Syria would come to an end, and new strategic threats—namely, preventing Iran from expanding the "Iranian Crescent"[328] that King Abdullah had warned of from Tehran to the Mediterranean—would emerge.[329] Long-term goals were becoming short-term objectives. Still, the battle for Mosul continued. Coalition forces pushed along a malleable front in Iraq and Syria. A force that once threatened to spread from the Euphrates and Tigris to the Atlantic Ocean was disintegrating in a storm of fire. But to the GID, there was still unfinished business on the agenda. There were two names remaining on the GID list.

Abu Khattab al-Rawi carried two dubious portfolios for the Islamic State. He was Abu Mohammed al-Adnani's deputy who inherited the spot as Caliphate media head after his boss was incinerated by coalition aircraft. He was also a master tinkerer and was responsible for the fledgling Islamic State drone air force. Islamic State drones, commercial-like flying devices fitted with explosive payloads, had proven to be most effective in targeting coalition forces during the battle for Mosul.

As al-Adnani's deputy, the Iraqi-born al-Rawi was a cheerleader. He was young—some believed he was no older than twenty-five—and he applauded all of his boss's ideas and actions with the exuberance of a staffer dreaming of a promotion. Al-Rawi made sure that al-Adnani's instructions were carried out to the letter. After Moaz was seized, it was al-Rawi who coordinated the production, assembled the photographic gear, helped to select the location, and worried about lighting and choreographing the executioners. Al-Rawi worked on the hostage video that the Raqqa Media Center produced and later disseminated to the Arab world. Al-Adnani's effectiveness as head of external operations was enhanced by his trusted lieutenant. A member of the Moaz Task Force likened al-Rawi to a hyperproductive office secretary. "When the boss bellowed," a GID officer commented, "al-Rawi made it happen."[330]

Al-Rawi had a particular disdain for Jordanians. Iraq and Jordan were connected by history and culture, but inherent jealousies and animosities existed on both sides of the frontier. Al-Rawi hated the role that Jordan had played in the international coalition and the fact that Moaz al-Kasasbeh's execution had failed to ignite jihadist passions inside the kingdom. Undeterred, al-Rawi was determined to strike Jordan. At dawn on June 21, 2016, a suicide bomber crashed his explosive-laden vehicle into a forward military post in the no-man's-land separating Jordan and Syria near the Rukban refugee center. Jordanian soldiers were cooking the morning meal for newly arrived refugees when the bomber detonated his payload. Six Jordanian soldiers were killed in the blast, and fourteen more were critically wounded.[331] The attack was carried out by a cell directly under al-Rawi's command. It was an impressive achievement for one so young.

As the Islamic State's fortunes soured on the battlefield, al-Rawi found shelter in his native Iraq. He shuttled between Ba'aj, a town in the westernmost district of Nineveh Province on the Syrian border,[332] and the town of al-Qa'im in Anbar Province; al-Qa'im, situated along the Euphrates River, was a notorious smuggling center connecting southern Iraq and Syria and was a transit hub for fighters traveling to and from the Islamic State battlefields.* Al-Rawi had been a fixture on the off-road trails that the local tribesmen used for evading customs officers since the days of Zarqawi.

The GID's human network of reliable sources was, perhaps, strongest in Anbar Province. The Iraqi Bedouin tribes were connected by blood and by marriage with their counterparts in Jordan, and many of the intelligence officers serving in the CTB came from those extended clans. "The tribe was everything," a retired Christian intelligence officer in the GID explained. "It was more important than the nation-state. Far more important than money. The bonds were even more important than religion."[333]

It was relatively easy for the Jordanian intelligence officers to meet with the tribal leaders in the province.† All meets were fraught with danger, but visitors were guests inside the tribal tent, and were protected. The Americans used to travel in heavily armored vehicles supported by heavily armed men. Such precautions were not needed inside the encampment.

Women were kept out of view. In most cases, the men did the cooking—and the cleaning. There were always young children curiously watching what transpired. The children were always eager observers every time the men from across the desert came calling in their shiny new vehicles.

Discussions between the Iraqis and the Jordanian intelligence officers were done in true Bedouin style: the slaughter of a sheep or goat, the two top-ranking men from both sides sitting next to each other in the tent and then a feast. The conversations were never private; one-on-one communications would have been an insult to the tribal leaders. The elders were to be trusted.

The old men—the sheiks—were savvy and wise in a way that

---

* In 2005, in fact, the US Marine Corps spearheaded a hard-fought eleven-day offensive in al-Qa'im against the hub that facilitated the back-and-forth travel of foreign volunteers to Zarqawi's network. The fighting was fierce and often hand-to-hand, but when the operation (code-named "Matador") concluded, over one hundred terrorists lay dead.

† Security considerations require that certain names and locations be altered.

westerners couldn't quite appreciate. These men survived Saddam, two American invasions, bloody battles with the Shiites and two Islamic insurgencies. They were survivors and pragmatic thinkers. They knew exactly what type of Islamic State person would be of interest to the GID. Few in the province knew what al-Rawi looked like, but they were certain they'd know how to find him. It was the Bedouin way: sometimes to honor their own way of life, it was essential that they betray others—especially if the others were outsiders or carried with them a foreign way of thought or worship.

It took months for the GID's friends in the desert to locate al-Rawi. The search was done subtly, with a nomadic tact that ensured al-Rawi's Emni bodyguards would not burrow their protectee further underground. Al-Rawi had a youthful energy that infuriated those who hunted him. He moved about hourly and with a small footprint. He masqueraded as a farmer and even as a woman.

In the desert, though, the Bedouin always persevere. Somehow al-Rawi was located. A pattern of his movements was established. Insiders became assets.

Details of exactly how the GID received word of al-Rawi's location remain classified—the precise manner how the intelligence was relayed will probably never be publicly released. But on May 26, 2017, Abu Khattab al-Rawi was killed in a coalition air strike against a media center in al-Qa'im.[334]

There was one man left that was marked for death.

Abdul-Hai al-Muhajir was the last name on the Moaz Task Force's assassination list. The Iraqi-born Islamic State commander was considered the most mysterious of the men hunted. The wali of Damascus, he was the Caliphate's southern front commander. He had urged al-Baghdadi to authorize the immolation of Moaz al-Kasasbeh in the hope of turning Jordan into the next war zone. Al-Muhajir had hoped to lead a black-banner-waving army to the Jordanian frontier and then march it all the way to Amman.

Al-Muhajir's final moments remain shrouded in mystery.

"The information on this terrorist was handed to the CIA," a GID officer named Saleem* explained, "and then they saw to it that he met his end." All that can be said about al-Muhajir's death was that it occurred sometime on September 17, 2017. It is believed that a drone incinerated the Toyota Land Cruiser he was driving one autumn morning in southern Syria.

As what remained of al-Muhajir's corpse burned in the fiery wreck of a vehicle decimated by a Hellfire antitank missile, the last embers of what was the Caliphate died out. Mosul had fallen to coalition-led Iraqi forces in the summer; Raqqa, the Caliphate's spiritual capital, was a month away from liberation by the SDF. Russia, the Assad loyalists, the Iranians and Hezbollah, and the Turks continue to fight over the future of Syria— emboldened by what can best be described as the United States unilaterally washing its hands of the region and its allies. The war against the Islamic State, though, was basically over. Al-Baghdadi's forces no longer held swaths of territory and controlled vast wealth. They were forced underground, like bandits.

When Moaz al-Kasasbeh fell into the hands of the Islamic State on December 24, 2014, Abu Bakr al-Baghdadi and his cohorts could have shown the young airman simple decency to send a message to the Arab and Muslim world that they could display compassion to one of their own. Instead, the Jordanian F-16 pilot suffered a barbaric death—a murder that horrified the world and propelled an alliance of covert warriors into decisive action. When Abu Mohammed al-Adnani ignited the accelerant that ultimately burned the son of Kerak to death, he thought he was igniting a spark that would rally the entire Muslim world behind him. All he really did was light the fuse of the Islamic State's destruction.

★ ★ ★ ★ ★

---

* A pseudonym.

POSTSCRIPT

Caution and Vigilance

THE DESERT GARRISON of American and Jordanian special operations units in al-Tanf was an instrumental facet of the coalition's effort to stem the southern stretch of the Iraqi desert from becoming a new white-hot front in the conflict. The joint teams mounted aggressive patrols to maintain a fifty-five-kilometer-wide security buffer against any and all threats or terrorist movements moving in and out of Syria and Iraq.

On November 16, 2017, a small American-Jordanian team based out of al-Tanf was on patrol outside of the base's gates when it encountered a convoy of vehicles ferrying a large number of men, women and children. The operators weren't too suspicious at first because, several miles away, the vehicles had passed through a checkpoint of what was known in the vernacular as PRFs, or pro-regime forces, namely Iranian and Hezbollah intelligence squads disguised as someone else. When the American and Jordanian operators attempted to ascertain the identities of the people in the long line of vehicles, they came under hellacious automatic weapons and RPG fire. Several gunmen emerged from the vehicles wearing suicide vests and attempted to get close enough to the Jordanian and American team members to detonate their payloads. The operators tried to fend off the attack while not harming the women and children who were being used as human shields. The Americans and Jordanians were also wary of instigating any contact with the militia forces nearby, who were itching for an incident that would justify hostile contact.

A pitched battle ensued, followed by a desperate close-quarter firefight. When the smoke cleared and the bodies were counted, the mixed special operations force had killed thirty Islamic State fighters and captured several more. None of the civilians were harmed.[335]

The Jordanian and American operators were decorated for their valor under fire two months later at a quiet ceremony held at the Special Forces Battalion headquarters in the outskirts of Amman. The men were commended for their quick thinking and courage. But they were back at al-Tanf later that evening. Although Mosul had fallen, as had Raqqa, the war went on: the Americans and Jordanians remain at al-Tanf as a hedge against Iranian power moving into the vacuum.

As much as wars change the Middle East, the region remains reliably dangerous.

On August 10, 2018, an Islamic State–inspired cell attacked a festival in a predominantly Christian town near the Jordanian capital, killing a police officer. When security forces, including members of the CTB-71, chased the suspects, a standoff ensued in a block of flats in the town of Salt. When the commandos stormed the location, the terrorists detonated their explosives, causing the building to collapse. Four operators were killed in the raid; all of the terrorists were killed.[336] The GID located guns, ammunition and explosives in a flat used by the terrorists.

Islamic State franchises are still fighting in North Africa, in Yemen and in Nigeria; they still pose a threat in Jordan, in the Persian Gulf and in parts of Europe. And even though the Caliphate was dissolved into almost nothing in Iraq and Syria, dangerous pockets remain. The Islamic State is still active in some of the Sunni villages in between Iraqi and Kurdish lines. They use night and fear as covers; they attack when the opportunity affords.[337] The reality on the ground contradicted the claim by US president Donald Trump that the war against the Islamic State was "100 percent over."[338] The president intended to bring all of the forces back home, including the special operations units assisting the Kurds and other allies, to make the statement that victory had finally come.

But in early August 2019, the Pentagon released a scathing indictment of US policy in Iraq and Syria, claiming that the "unilateral scaling down of American ground troops and special operations forces from the area has inadvertently aided the Islamic State's regrouping in Syria and Iraq, and helped it transition from a territory-holding force to an insurgency."[339]

Robin Wright, the award-winning journalist and author and joint fellow at the US Institute of Peace and the Woodrow Wilson International Center, summed it up best when, in an essay for the *New Yorker*, she wrote, "History will record that the Islamic State caliphate—a bizarre pseudo-state founded on illusory goals, created by a global horde of jihadis, and forced with perverted viciousness—survived for three years, three months and some eighteen days."[340] The history of conflict in the Middle East involving religion and real estate is as old as time itself. The Islamic State will be remembered in the larger scheme of history as a blip, an aberration of human madness that was avoidable in so many ways—created mostly by unsolvable enmities and outside politicians who never understood what made the area tick. In the end, it took spies, soldiers and the combined operations of allied nations to correct the blunders and miscues of the political leaders who let the genie out of the bottle.

But world leaders must remain cautious and on guard. The Islamic State has been defeated, but it is far from dead. The international coalition that was assembled following the fall of

Mosul succeeded in stemming the advance, yet pockets have been allowed—by fate or by folly—to survive. Containing the Islamic State and ultimately killing it—by war, by education and by political leadership—will require courage and craftiness.

Jordan's King Abdullah initiated an attempt to contain the trip wires of what ultimately led to the creation of the Islamic State. Known as the Aqaba Process, the effort is a low-key attempt, far from the prying eyes of news organizations, to honestly understand what are the root causes of Islamic extremism, and what measures the region's—and the world's—leaders can initiate in order to make sure that what led to al-Baghdadi's rise and what prompted so many volunteers to leave their homes and families are stopped before they begin. The meetings between the world leaders involved are meant to be relaxed and personal: no jackets and ties, no entourage of staffers, and none of the formalities that often stifle bilateral discussions.

The entire process was designed to see how global leaders could apply a holistic approach to all forms of extremism, radicalization and counterterrorism within a joint international effort, based on close coordination and consultations that cover military, security and ideology aspects.[341] The concept was global and meant to ensure that the nations of the world could prevent the sudden eruption of ultraviolent extremism before they had to go to war. The success of this process requires steadfast commitment and preemptively eliminating dangerous trends before they become dire threats. The spies will need to remain vigilant. Soldiers will need to respond to danger close and far, requiring many to make the ultimate sacrifice. The alliance between nations—and between the spy services—will provide an invisible defense shield against any attempts to relaunch such destructive and ambitious forces.

And it is because of the strategic imperative of alliances—local, regional and global—that the October 2019 decision by US President Donald Trump to withdraw the small contingent of American special operations forces from northern Syria, in areas that Kurdish forces liberated from ISIS, came as such a jarring surprise. Although the American president had forewarned of his desire to remove the American Special Forces presence in Syria, the move had immediate geopolitical implications. The vacuum left by the retreating special operations units was quickly filled by Turkey, Russia and Syria; the Syrian Kurds, who had already sacrificed so much in the war against the Caliphate, losing over 11,000 dead and scores more wounded, were once again forced to withstand aerial bombardments and artillery barrages in a fight for their survival. The alliance that had defeated ISIS militarily was reassigned and redesigned, providing the so-called Islamic State with a possible lifeline after it had been left for dead on the battlefield. Even after Operation Kayla Mueller, the October 26-27, 2019 JSOC raid that killed ISIS founder Abu Bakr al-Baghdadi in Idlib Governorate, for Western intelligence service directors, military commanders, and the politicians who try so hard to maneuver through the Middle East, the concern was that a strategic and tactical alliance that worked based on trust and a common enemy might never be able to rise again when needed.

The Caliphate no longer exists on the map, but the evil that prompted the beheadings of so many and the immolation of a young Jordanian pilot—and the chaos that allowed the barbarity to spread—still flourishes in the hearts and minds of men who equate faith with carnage.

ACKNOWLEDGMENTS

A BOOK OF this kind would never have been possible without the assistance, friendship and generosity of a great many people who, because of the work they do and the flags they serve, can never be acknowledged in a public forum. These men and women fight in the shadows, concealed from view. They must remain hidden if they are going to be successful—and safe—in executing their duties, and for the sake of all of us, it's in our collective best interest that these warriors, spies, security specialists and decision makers remain safe and unencumbered by the public outing. I received enormous assistance from these individuals who spoke Arabic, English, French, German, Kurdish and more. They believed in this project because they felt it important that the story of the alliance between spies and soldiers from like-minded nations, especially in the context of a single story in the war against terror, was important to be told. Their role in the fight had to remain anonymous. I am very grateful to these brave men and women and I thank them for their assistance, for their guidance and for their service.

I would like to thank many members of the US intelligence fraternity, the US special operations community and members of the State Department—past and present—for their assistance in assembling the book. Many of these remarkable men and women have spent lifetimes away from their families serving their country in places that became the front lines in the war against the Islamic State. Alas, as many of these people are still in uniform and still carry "creds," they must be thanked in quiet. In this day and age of free-flowing information, those who work for the respective governments in the free world have become very wary of attributable free-floating information. Officials, even those tasked with disseminating updates from the field of battle, are reticent to attach their names to news and insight of any importance—the world wide web and social media have made everyone paranoid. As a result, much of the information I received for this book was on the condition of anonymity. I am grateful to those out there who helped and risked so much to make sure I got the story right. You all know who you are. Thanks!

Jordan is a very special place where history and future merge into a collective mosaic. The country is tribal, yet it welcomes people from all walks of life with open arms. It is a nation that has fought gallantly to protect its interests, yet it has provided shelter and security for refugees from all around the region in need of a safe haven. Jordan is a nation that knows the peril of terror and, mostly, the steps that need to be taken to fight it. It has been my great luck to be working in Jordan—with the country's tip-of-the-spear elite special operations units—since 1994, and I am grateful to King Abdullah for his generosity and kindness.

I had the chance to work with many members from the GID and the Jordan Armed Forces, and I thank them for their friendship and kindness—for the obvious reasons they must remain nameless.

I would like to thank the staffers at the Zaatari refugee camp for doing God's work, and I'd like to thank Dr. Musa Shteiwi at the Center of Strategic Studies in the University of Jordan for providing me with some of his expertise on the radicalization process involved in turning middle-class sons into suicidal terrorists.

I've made many friends in Jordan over the years and they all played a critical role in the writing of this book. I'd like to thank my dear friend and partner Ayman Masri for his loyal guidance and support. I'd like to thank the members of the Amman Gang—Dr. Saleem, Khaled and Ghassan—for always being ready for a smile, a good time, and some deep and very insightful discussion and political dissection in some of the coolest Amman eateries imaginable.

I would also like to offer a very heartfelt and special thanks to two dear and amazing people and one of the coolest power couples imaginable: Ranya Kadri and Adie As'ad. If you are friends with Ranya, then you are friends with everyone in Jordan. She is a source of endless knowledge and deep perspective in a region that is all too often defined by misunderstanding. Ranya has worked for some of the most prestigious newspapers and news outlets in the world, and many of this planet's top journalists—people with Pulitzer Prizes and films to their credit—have churned out award-winning manuscripts and copy from "Casa de Ranya and Adie" in a guest room overlooking the hills of Amman. Ranya and Adie welcomed me into their home and into their family—I can't imagine a bigger honor.

This project provided me with the opportunity to travel to Iraq and to the Kurdistan Regional Government. The Kurds are terrific people—infectiously friendly and proud of their culture, their penchant for survival and their courage in this last conflict. I'd like to thank Alex Asbery from the KRG office in Washington, DC, for helping me make the travel happen, and to Omed Ahmed for making sure the visit was perfectly planned. I'd like to thank Peshmerga brigadier Omer Ismail for his insight and hospitality, Lieutenant General Jamal Mohammed, the Peshmerga chief of staff, and Peshmerga deputy minister Sarbast Lazgin Sendury for taking the time out of their busy schedules to provide me with an in-depth briefing on the war that was and the conflicts likely to come. A special thanks to Major General Sirwan Barzani and his assistant "M" for a memorable tour of the front lines near Makhmur.

I would also to offer a special word of thanks to the protective detail from the US State Department Diplomatic Security Service at the US Consulate in Erbil. DSS is without doubt America's finest!

Finally, a special thanks to my friend "Brian," a soldier and a gentleman, whose last name is withheld for the obvious reasons. "Brian" was a generous and perfect travel colleague to Erbil and beyond.

This book was just an idea when I approached Jim Hornfischer, my terrific literary agent. I pitched it to him and turned it into a proposal, and then Jim did his magic and helped me to transform my ideas into a coherent and marketable document. There is no better guide through the minefields of publishing than Jim. I am really lucky to have him on my side. I would like to thank my editor, Peter Joseph, for making sure that this book had a home, and for being such a pleasure to work with. Peter and his team at Hanover Square Press are awesome: all authors should be so lucky. I'd like to thank Fred Burton, whom I've coauthored several books with and who is a dependable friend who was crucial in helping me fill in those pesky gaps when it came to the sensitive world of government agencies and espionage. A special thanks to my dear friend and colleague Ziv Koren. Ziv is one of the world's top combat photographers who has seen his fair amount of danger. He has been inside terrorist tunnels and hostage events, and he was kind enough to take time away from his frenetic globetrotting schedule to take a professional author photo of me. Finally, a special thanks to Dr. Ariela Frieder and Felipe Dorregaray—their kindness, generosity and friendship were an essential piece of the book-writing process.

Writers are a frustrating breed—we are self-absorbed, we keep odd hours, we tend to be overwhelmed, and our minds and—when we travel—our bodies are thousands of miles away. We don't work neatly; we aren't easy to live with. So, I would like to thank my wife and three children for their patience and—most of all—their tolerance. My family inspires me and makes the hard work and the sometimes challenging realities of writing for a living worth it all. I am truly blessed.

Samuel M. Katz
September 2019

ENDNOTES

1 Interview, Captain Sa'ab (a pseudonym), No. 1 Squadron pilot, Muwaf-faq Salti Air Base, September 16, 2018.

2 Fayek S. Hourani, *Daily Bread for Your Mind and Soul: A Handbook of Transcultural Proverb and Sayings* (Bloomington, Indiana: Xlibris Corpo-rations, 2012), p. 187.

3 Mark Landler, "Rice to Replace Donilon in the Top National Security Post," *New York Times*, June 5, 2013.

4 Manuel Roig-Franzia, "Susan Rice: Not Your Typical Diplomat," *Wash-ington Post*, November 29, 2012.

5 Charles R. Lister, *The Syrian Jihad: The Evolution of an Insurgency* (Lon-don: C. Hurst and Company, 2015,), 11.

6 Lister, *Syrian Jihad*, 12.

7 "Islamist Group Claims Syria Bombs 'to Avenge Sunnis,'" *Al-Arabiya News*, March 21, 2012.

8 Rukmini Callimachi, "ISIS Enshrines a Theology of Rape," *New York Times*, August 13, 2015.

9 Mark MacKinnon, "The Graffiti Kids Who Sparked The Syrian War," *Globe and Mail*, December 2, 2016.

10 Jeremy Bender, "Jordan's Special Forces Are Some of the Best in the Middle East," *Business Insider*, April 22, 2016.

11 Interview, Amman, March 28, 2019.

12 Babak Dehghanpisheh, "Iran's elite Guards fighting in Iraq to push back Islamic State," Reuters, August 3, 2014.

13 Interview, King Abdullah, Amman, September 17, 2018.

14 Nabih Bulos, W. J. Hennigan, and Brian Bennett, "In Syria, Militias Armed by the Pentagon Fight Those Armed by the CIA," *Los Angeles Times*, March 27, 2016.

15 Mark Landler, "Rice Offers a More Modest Strategy for Mideast," *New York Times*, October 26, 2013.

16 Associated Press, "Retired Gen. Norman Schwarzkopf Dies," *Politico*, December 27, 2012.

17 David D. Kirkpatrick, "Graft Hobbles Iraq's Military in Fighting ISIS," *New York Times*, November 23, 2014.

18 Ned Parker, Isabel Coles, and Raheem Salman, "Special Report—How Mosul Fell: An Iraqi General Disputes Baghdad's Story," Reuters UK, October 14, 2014.

19 Andrew Slater, "The Monster of Mosul: How a Sadistic General Helped ISIS Win," *Daily Beast*, June 19, 2014.

20 Mark Kurkis, "Last Chance at an Iraqi Prosecution?" *Time*, June 16, 2008.

21 Interview, Washington, DC, July 19, 2018.

22 Interview, Majd Holbi, July 8, 2018.

23 Michael Weiss and Hassan Hassan, *ISIS: Inside the Army of Terror* (New York: Regan Arts Books, 2015), 120.

24 Martin Chulov, "ISIS: The Inside Story," *Guardian*, December 11, 2014.

25 Terrence McCoy, "How the Islamic State Evolved in an American Prison," *Washington Post*, November 4, 2014.

26 Associated Press, "Iraq: Hundreds Escape from Abu Ghraib Jail," *Guardian*, July 22, 2013.

27 Rose Troup Buchanan, "Daesh: What Does the Word Mean? Is David Cameron's Use of It Significant?," *Independent*, December 2, 2015.

28 Rukmini Callimachi, "ISIS Enshrines a Theology of Rape," *New York Times*, August 13, 2015.

29 Interview, Majd Holbi, July 8, 2018.

30 Ned Parker, Isabel Coles, and Raheem Salman, "Special Report—How Mosul Fell: An Iraqi General Disputes Baghdad's Story," Reuters UK, October 14, 2014.

31 Interview, Majd Holbi, July 8, 2018.

32 https://www.youtube.com/watch?v=2OuxgHojAis&t=262s

33 Ned Parker, Isabel Coles, and Raheem Salman, "Special Report—How Mosul Fell: An Iraqi General Disputes Baghdad's Story," Reuters UK, October 14, 2014.

34 https://www.youtube.com/watch?reload=9&v=ECJqr3HN6_Q

35 Human Rights Watch (@hrw), "Iraq: ISIS Executed Hundreds of Prison Inmates About 600 Shias Killed in Desert During Mosul Capture," Twitter, October 31, 2014, 9:55 a.m., https://twitter.com/hrw/status/528198480251330561.

36 "Sunni Rebels Declare New 'Islamic Caliphate,'" Al Jazeera, June 30, 2014, https://www.aljazeera.com/news/middleeast/2014/06/isil-declares-new-islamic-caliphate-201462917326669749.html.

37 Hannah Strange, "Islamic State Leader Abu Bakr al-Baghdadi Addresses Muslims in Mosul," *Telegraph*, July 5, 2014.

38 John le Carré, interview by Ramona Koval, The Book Show, Australian Broadcasting Commission Radio National, November 19, 2008.

39 Jack O'Connell with Vernon Loeb, *King's Counsel: A Memoir of War, Espionage, and Diplomacy in the Middle East* (New York: W.W. Norton and Company, 2011), 4.

40 https://gid.gov.jo/gid-info/about-g-i-d/

41 Christopher Dobson, *Black September* (New York: Macmillan, 1974), 1.

42 Emily Cadei, "Eyes and Ears in the Arab World," Ozy, March 3, 2015.

43 Interview, Amman, September 15, 2018.

44 A conversation with John le Carré: https://www.youtube.com/watch?v=89FHIGL3N54&t=193s.

45 Joby Warrick, "Jordan Emerges as Key CIA Counterterrorism Ally," *Washington Post*, January 4, 2010.

46 https://www.youtube.com/watch?v=iafsOfRasW8&feature=youtu.be

47 Interview, Amman, September 15, 2018.

48 Gary Gambill, "Abu Musab al-Zarqawi: A Biographical Sketch," Terrorism Monitor, Volume 2, Issue 24, December 16, 2004 (the Jamestown Foundation).

49 Craig Whitlock, "Abu Musab Zarqawi's Biography," *Washington Post*, June 8, 2006.

50 Joby Warrick, *Black Flags: The Rise of ISIS* (New York: First Anchor Books, 2016), 44.

51 http://www.washingtonpost.com/wp-dyn/content/article/2006/06/08/AR2006060800299_3.html?nav=rss_world/africa&noredirect=on

52 Interview, New York City, November 1, 2018.

53 Cameron Reed, "America's Best Partner in Middle East HUMINT Needs Help," Defense One, June 22, 2017, https://www.defenseone.com/ideas/2017/06/americas-best-partner-middle-east-humint-needs-help/138889/.

54 Interview, September 15, 2018.

55 Timothy J. Burger, Viveca Novak, and Elaine Shannon, "Threat Analysis: Decoding the Chatter," *Time*, October 31, 2010.

56 "John O. Brennan: Deputy National Security Advisor for Homeland Security and Counterterrorism," Washington Politics, *Washington Post*, January 8, 2013.

57 Interview, Amman, September 17, 2018.

58 "Obama Warns Syria Not to Cross 'Red Line,'" CNN, August 21, 2012, https://www.cnn.com/2012/08/20/world/meast/syria-unrest/index.html.

59 Interview, September 15, 2018.

60 Patrice Taddonio, "WATCH: Pentagon Had No Plan for When ISIS Took Mosul, Dempsey Says," *Frontline*, May 26, 2015.

61 https://www.pbs.org/wgbh/frontline/film/obama-at-war/transcript/

62 Ben Rhodes, "Inside the White House During the Syrian 'Red Line' Crisis," *Atlantic*, June 3, 2018.

63 Rhodes, "Inside the White House."

64 https://www.inherentresolve.mil/Portals/14/Documents/Mission/HISTORY_17OCT2014-JUL2017.pdf?ver=2017-07-22-095806-793

65 Sebastian Vega, "Former Mossad Director Speaks on Campus about Islamic State Group," *Daily Texan*, October 15, 2014, https://www.dailytexanonline.com/2014/10/15/former-israel-defense-forces-officer-speaks-on-campus-about-islamic-state-group.

66 Julie Hirschfeld Davis and Steven Erlanger, "NATO Weighs Rapid Response Force for Eastern Europe," *New York Times*, September 1, 2014.

67 Kim Hjelmgaard, "NATO Summit 'Most Important' Since Fall of Berlin Wall," *USA Today*, August 31, 2014.

68 Suha Ma'ayeh, "How Jordan Got Pulled into the Fight against ISIS," *Time*, February 26, 2015.

69 Interview, Professor Musa Shteiwi, director of the Center for Strategic Studies at the University of Jordan, Amman, September 16, 2018.

70 "Wales Summit Declaration," Issued by the Heads of State and Government participating in the meeting of the North Atlantic Council in Wales, September 5, 2014, https://www.nato.int/cps/ru/natohq/official_texts_112964.htm?selectedLocale=en

71 Julie Hirschfeld Davis, "Obama and Cameron Call on NATO to Confront ISIS," *New York Times*, September 4, 2014.

72 Interview, King Abdullah, Amman, September 17, 2018.

73 https://media.defense.gov/2017/Apr/06/2001728012/-1/-1/0/B_0106_WESTENHOFF_MILITARY_AIRPOWER.PDF

74 https://dod.defense.gov/OIR/Airstrikes/

75 Interview, Brigadier General Ghassan (pseudonym), commanding officer of Muwaffaq Salti Air Base, September 16, 2018.

76 Interview, No. 1 Squadron commander Lieutenant Colonel Ahmed (a pseudonym), Muwaffaq Salti Air Base, September 16, 2018.

77 Interview, Lieutenant Colonel Ahmed, September 16, 2018.

78 Interview, Muwaffaq Salti Air Base, September 16, 2018.

79 Interview, Muwaffaq Salti Air Base, September 16, 2018.

80 US Government Accountability Office, *Defense Acquisitions: Assessments Needed to Address V-22 Aircraft Operational and Cost Concerns to Define Future Investments*, GAO-09-482, May 11, 2009, https://www.gao.gov/products/GAO-09-482.

81 Jeff Schogol, "A-10 Attacking Islamic State Targets in Iraq," *Military Times*, December 18, 2014.

82 Interview, Brigadier General Ghassan, RJAF headquarters, Marka Air Base, March 27, 2019.

83 https://www.youtube.com/watch?v=pjdP2t7Uarw

84 John McCain, speech to the American Red Cross "Promise of Humanity Conference," May 6, 1999.

85 Interview, No. 1 Squadron commander Lieutenant Colonel Ali (a pseudonym), Muwaffaq Salti Air Base, September 16, 2018.

86 Interview, King Abdullah, Amman, September 17, 2018.

87 Interview, King Abdullah, Amman, September 17, 2018.

88 Interview, Amman, February 22, 2019.

89 Dan Deluce and Ruby Mellon, "Trump Dumped U.S. Ambassador to Amman at Request of King," *Foreign Affairs*, August 31, 2017.

90 Interview, Amman, September 17, 2018.

91 George Packer, "Exporting Jihad," *New Yorker*, March 28, 2016.

92 https://www.brookings.edu/wp-content/uploads/2014/12/en_whos_who.pdf

93 Tim Walker, "James Foley's Final Days: Kept in a Cell with 18 Other Captives, Beaten, Starved, Waterboarded—and Apparently Abandoned by His Government," *Independent*, October 27, 2014.

94 Tim Walker and Rob Hastings, "Steven Sotloff 'Beheading': British Captive Shown in ISIS Video Which Claims to Show Death of Second US Journalist," *Independent*, September 2, 2014.

95 Interview, New York, October 1, 2014.

96 Greg Miller and Souad Mekhennet, "Inside the Surreal World of the Islamic State's Propaganda Machine," *Washington Post*, November 20, 2015.

97 https://www.terrorism-info.org.il/app/uploads/2019/02/E_285_18.pdf

98 https://www.memri.org/tv/people-behind-isis-media-system-speak-al-arabiya-tv-documentary/transcript

99 https://www.nytimes.com/video/multimedia/100000003418172/father-of-captured-jordanian-pilot-pleas-for-sons-welfare.html

100 Ben Hubbard, "ISIS Militants Capture Jordanian Fighter Pilot in Syria," *New York Times*, December 24, 2014.

101 Julian Droogan and Shane Peattie, "Dabiq: The Changing Face of Jihad Propaganda," *Australian Outlook*, Australian Institute for International Affairs, October 3, 2017.

102 Andrew V. Pestano, "Islamic State Publishes Interview with Captured Pilot, Asks Twitter Followers How to Kill Him," UPI, December 30, 2014.

103 Suleiman al-Khalidi, "ISIS Releases Apparent Interview with Captured Jordanian Pilot," *Huffington Post*, December 30, 2014.

104 "Jordanian Pilot Captured by ISIS Says Militants 'Will Kill Me' in Magazine Interview," Agency, *Independent*, December 30, 2014.

105 Terrence McCoy, "Why the Islamic State Wants to Free Forgotten Suicide Bomber Sajida Rishawi," *Washington Post*, January 26, 2015.

106 Michael D. Shear and Eric Schmitt, "In Raid to Save Foley and Other Hostages, U.S. Found None," *New York Times*, August 20, 2014.

107 Adam Goldman and Karen DeYoung, "U.S. Staged Secret Operation into Syria in Failed Bid to Rescue Americans," *Washington Post*, August 20, 2014.

108 Interview, Lieutenant Colonel Ahmed, Muwaffaq Salti Air Base, September 16, 2018.

109 Interview, Lieutenant Colonel Ahmed, September 16, 2018.

110 Interview, Lieutenant Colonel Ahmed, September 16, 2018.

111 Interview, Amman, March 27, 2019.

112 Rebecca Tan, "Terrorists' Love for Telegram, Explained," *Vox*, June 30, 2017, https://www.vox.com/world/2017/6/30/15886506/terrorism-isis-telegram-social-media-russia-pavel-durov-twitter.

113 Richard Barrett, *The Islamic State* (New York: The Soufan Group, November 2014).

114 Interview, October 22, 2017.

115 Tara John, "Everything We Know about ISIS Spy Chief Abu Moham-mad al-Adnani," *Time*, August 30, 2016.

116 Interview, Dr. M., GID headquarters, Amman, October 27, 2017.

117 https://www.memri.org/tv/people-behind-isis-media-system-speak-al-arabiya-tv-documentary/transcript

118 Interview, GID headquarters, Amman, March 27, 2019.

119 Rod Nordland and Ranya Kadri, "All-but-Forgotten Prisoner in Jor-dan Is at Center of Swap Demand by ISIS," *New York Times*, January 28, 2015.

120 Luke Baker, "Islamic State Threatens Two Japanese Captives in Video," Reuters, January 20, 2015.

121 Reiji Yoshida, "Islamic State Threatens to Kill Two Japanese Hos-tages," *Japan Times*, January 20, 2015.

122 Simon Engler, "The U.S. Does Negotiate with Terrorists," *Foreign Policy*, June 3, 2014.

123 Interview, GID headquarters, Amman, September 16, 2018.

124 Rod Nordland, "Jordan Hostage Crisis May Hurt U.S. Ties," *New York Times*, January 27, 2015.

125 Nordland, "Jordan Hostage Crisis."

126 Interview, Washington, DC, January 14, 2019.

127 Interview, King Abdullah, Amman, September 17, 2018.

128 Hamza Hendawi, Qassim Abdul-Zahra, and Bassem Mroue, "ISIS Has 'Special Forces,'" *Business Insider*, July 8, 2015, https://www.business-insider.com/isis-has-special-forces-2015-7.

129 Nadeem Badsha, "SAS 'Saved Hundreds of Lives' in Battle with Is-lamic State for Kobani," *Express*, October 26, 2014.

130 Tom Rogan, "Who Dares Wins: How American and British Special Forces Smashed the Islamic State," *Washington Examiner*, December 20, 2018.

131 http://www.warfare.today/2018/05/09/news-story-lieutenant-general-mark-carleton-smith-appointed-new-chief-of-the-general-staff/

132 Anne Barnard and Karam Shoumali, "Kurd Militia Says ISIS Is Expelled from Kobani," *New York Times*, January 26, 2015.

133 Tuqa Nusairat, "Jordan's Pilot and the War on ISIS," Atlantic Council, February 3, 2015.

134 Heather Saul, "ISIS Films Reactions of Cheering Crowds Watching Jordanian Pilot Burned Alive on Screens in Raqqa," *Independent*, February 4, 2015.

135 Interview, Amman, January 30, 2019.

136 Hearing on the U.S. Role and Strategy in the Middle East, before the Senate Foreign Relations Committee, 114th Cong. (October 28, 2015) (statement of General John R. Allen, USMC [Ret.], Special Presidential Envoy for the Global Coalition to Counter ISIL).

137 Interview, Muwaffaq Salti Air Base, September 16, 2018.

138 Interview, Muwaffaq Salti Air Base, September 16, 2018.

139 Interview, Muwaffaq Salti Air Base, September 16, 2018.

140 Joby Warrick, "The King Demanded Vengeance and 'Zarqawi's Woman' Was Sent to the Gallows," *Washington Post*, September 26, 2015.

141 Interview, Amman, October 28, 2017.

142 Interview, September 16, 2018.

143 Joby Warrick, "The King Demanded Vengeance and 'Zarqawi's Woman' Was Sent to the Gallows," *Washington Post*, September 26, 2015.

144 Helene Cooper, "United Arab Emirates, Key U.S. Ally in ISIS Effort, Disengaged in December," *New York Times*, February 3, 2005.

145 Ian Black, "UAE Halted ISIS Air Attacks after Pilot Capture," *Guardian*, February 4, 2015.

146 Interview, Marka, September 16, 2018.

147 Joby Warrick, "The King Demanded Vengeance and 'Zarqawi's Woman' Was Sent to the Gallows," *Washington Post*, September 26, 2015.

148 Interview, King Abdullah, September 17, 2018.

149 "Jordan Executes Prisoners after ISIL Murder of Pilot," Al Jazeera, February 4, 2015, https://www.aljazeera.com/news/2015/02/jordan-executes-prisoners-isil-murder-pilot-150204060856804.html.

150 Interview, Amman, September 15, 2018.

151 Joby Warrick, "The King Demanded Vengeance and 'Zarqawi's Woman' Was Sent to the Gallows," *Washington Post*, September 26, 2015.

152 Martin Chulov, "Jordanians Turn Their Minds to Revenge After ISIS Killing of Pilot," *Guardian*, February 4, 2015.

153 Interview, Muwaffaq Salti Air Base, September 16, 2018.

154 https://www.cnn.com/2015/02/04/world/isis-jordan-reaction/index.html

155 https://kutv.com/news/nation-world/jordan-strikes-back-at-isis-after-pilot39s-killing

156 James Rush, "Annihilate ISIS: Father of Jordanian Pilot Burned to Death by Militants Calls for Revenge as Murder Sparks Outrage Across Middle East," *Belfast Telegraph*, February 4, 2015.

157 William Booth and Taylor Luck, "Jordan Hits Islamic State with Airstrikes as King Visits Family of Pilot Burned Alive," *Washington Post*, February 5, 2015.

158 Interview, GID headquarters, Amman, September 16, 2018.

159 Interview, Tel Aviv, January 7, 2019.

160 "Islamic State: Jordan's King Abdullah Vows 'Severe Response' to IS," BBC, February 4, 2015.

161 Interview, King Abdullah, Amman, September 17, 2018.

162 https://www.brainyquote.com/quotes/t_e_lawrence_789606

163 https://www.state.gov/discoverdiplomacy/docs/208086.htm

164 Interview, Amman, June 14, 2019.

165 Interview, Amman, February 26, 2019.

166 Interview, New York, February 13, 2019.

167 Interview, New York, February 13, 2019.

168 Shane Harris and Tim Mak, "The Most Gruesome Moments in the CIA 'Torture Report,'" Daily Beast, September 12, 2014, https://www.thedailybeast.com/the-most-gruesome-moments-in-the-cia-torture-report.

169 Interview, February 12, 2019.

170 Matt Schudel, "Charles McCarry, CIA Officer Who Became a Pre-Eminent Spy Novelist, Dies at 88," Washington Post, February 28, 2019.

171 Greg Miller and Julie Tate, "CIA's Global Response Staff Emerging from Shadows after Incidents in Libya and Pakistan," Washington Post, December 26, 2012.

172 Interview, New York, February 13, 2019.

173 Michael Field, "NZ Way Down the WikiLeaks Queue," Stuff, December 2, 2010.

174 Interview with retired state department agent, Texas, April 13, 2019.

175 Interview, New York, January 20, 2019.

176 Interview, Salhub, Jordan, June 13, 2019.

177 https://www.youtube.com/watch?v=tAvODK3er7Y

178 Allan Cullison, "Meet the Rebel Commander in Syria That Assad, Russia and the U.S. All Fear," *Wall Street Journal*, November 19, 2013.

179 https://www.youtube.com/watch?v=tAvODK3er7Y

180 https://www.treasury.gov/press-center/press-releases/Pages/jl2651.aspx

181 Bill Roggio, "Jordanian Cleric Extols Jihad at Funeral of AQAP Fighter," *FDD's Long War Journal*, June 29, 2017.

182 Feras Hanoush, "Who Will Succeed Al-Adnani?" Atlantic Council, September 20, 2016.

183 Bill Roggio, "Jordanian Cleric Extols Jihad at Funeral of AQAP Fighter," *FDD's Long War Journal*, June 29, 2017.

184 Interview, Amman, June 13, 2019.

185 Interview, Amman, March 26, 2019.

186 Ken Silverstein, "U.S., Jordan Forge Closer Ties in Covert War on Terrorism," *Los Angeles Times*, November 11, 2005.

187 Interview, Amman, March 26, 2019.

188 Ken Silverstein, "U.S., Jordan Forge Closer Ties in Covert War on Terrorism," *Los Angeles Times*, November 11, 2005.

189 Joshua Sinai, "How the Genocidal Successor to al Qaeda Persists," *Washington Times*, May 12, 2019.

190 https://www.indy100.com/article/isis-in-five-charts--ZyEuJ0Z4Fx

191 Interview, Professor Musa Shteiwi, Center for Strategic Studies, University of Amman, September 16, 2018.

192 Interview, Tel Aviv, May 19, 2019.

193 Interview, Zaatari refugee camp, September 14, 2018.

194 Inna Lazareva, "Israeli Doctor Treating Syrians Says Snipers Deliberately Shoot Children in the Spine," *Telegraph*, March 14, 2014.

195 https://reliefweb.int/sites/reliefweb.int/files/resources/2018.02.04FACTSHEET-ZaatariRefugeeCampFEB2018.pdf

196 https://data2.unhcr.org/en/situations/syria

197 *The Spymasters: CIA in the Crosshairs*, directed by Gédéon Naudet and Jules Naudet, written by Chris Whipple (2015, Showtime).

198 Alan Feuer, "A Nation at War: Jordan; Iraqi Agents Held in Plot to Poison Water Supply," *New York Times*, April 2, 2003.

199 Interview, Amman, March 27, 2019.

200 Borzou Daragahi and Josh Meyer, "'We Knew Him': Jordanian Spies Infiltrated Iraq to Find Zarqawi," *Los Angeles Times*, June 13, 2006.

201 Daragahi and Meyer, "We Knew Him."

202 Hassan Abu Haniyeh, *Daesh's Organizational Structure* (Mecca: Al Jazeera Centre for Studies, December 4, 2014), http://studies.al-jazeera.net/en/dossiers/decipheringdaeshoriginsimpactandfuture/2014/12/201412395930929444.html.

203 Christoph Reuter, "Secret Files Reveal the Structure of Islamic State," *Der Spiegel*, April 18, 2015.

204 Joseph Sassoon, "The East German Ministry for State Security and Iraq, 1968–1989," *Journal of Cold War Studies* 16, no. 1 (Winter 2014): 4–23.

205 Ann Speckhard and Ahmet S. Yayla, "The ISIS Emni: Origins and Inner Workings of ISIS's Intelligence Apparatus," *Perspectives on Terrorism* 11, no. 1 (February 2017).

206 Speckhard and Yayla, "ISIS Emni."

207 Interview, Amman, March 27, 2019.

208 Ann Speckhard and Ahmet S. Yayla, "The ISIS Emni: The Origins and Inner Workings of ISIS's Intelligence Apparatus," *Perspectives on Terrorism* 11, no. 1 (February 2017).

209 Speckhard and Yayla, "ISIS Emni."

210 Rukmini Callimachi, "ISIS Enshrines a Theology of Rape," *New York Times*, August 13, 2015.

211 Interview, Amman, March 28, 2019.

212 Christoph Reuter, "Secret Files Reveal the Structure of Islamic State," *Der Spiegel*, April 18, 2015.

213 Vera Mironova, Ekaterina Sergatskova, and Karam Alhamad, "ISIS' Intelligence Service Refuses to Die," *Foreign Affairs*, November 22, 2017.

214 Mironova, Sergatskova, and Alhamad, "ISIS' Intelligence Service."

215 Interview, Amman, GID headquarters, March 26, 2019.

216 Nabih Bulos, W. J. Hennigan, and Brian Bennett, "In Syria, Militias Armed by the Pentagon Fight Those Armed by the CIA," *Los Angeles Times*, March 27, 2016.

217 https://www.youtube.com/watch?v=HNXQS-lik64

218 https://web.archive.org/web/20111028221528/http://usaid.gov/iraq/accomplishments/USAIDAccomplishments2003-2009_October_2009.pdf

219 https://rewardsforjustice.net/english/mir_aimal_kansi.html

220 https://rewardsforjustice.net/english/uday_hussein.html and https://rewardsforjustice.net/english/qusay_hussein.html

221 Interview, New York, February 11, 2019.

222 Eric Schmitt, "A Raid on ISIS Yields a Trove of Intelligence," *New York Times*, June 8, 2015.

223 https://www.counterextremism.com/extremists/abu-sayyaf

224 Ken Dilanian, "Family Says Daughter Raped Repeatedly while Held by IS," Associated Press, August 15, 2015, https://apnews.com/4796751e2597413398e02da1d2fdbb4c.

225 Duygu Guvenc and Mitchell Prothero, "Turkey Vows to Fight Islamic State, Calls It 'Primary Threat,'" *Charlotte Observer*, July 24, 2015, https://www.charlotteobserver.com/news/nation-world/world/article28644250.html.

226 Guvenc and Prothero, "Turkey Vows to Fight."

227 Ahmet S. Yayla, Colin P. Clarke, "Ankara Claims to Oppose the Islamic State. Its Actions Suggest Otherwise," Foreign Policy, April 12, 2018

228 https://www.cia.gov/library/readingroom/docs/CIA-RDP85T00353R000100290001-4.pdf

229 Interview, King Abdullah Air Base, Amman, March 26, 2019.

230 Laila Bassam and Tom Perry, "How Iranian General Plotted Out Syrian Assault in Moscow," Reuters, October 6, 2015.

231 https://www.youtube.com/watch?v=2Iu_4GMKEXM

232 Peter Baker and Neil MacFarquhar, "Obama Sees Russia Failing in Syria Effort," *New York Times*, October 2, 2015.

233 Andrew S. Weiss and Nicole Ng, "Collision Avoidance: The Lessons of U.S. and Russian Operations in Syria" (paper, Carnegie Endowment for International Peace, March 20, 2019).

234 Mike Giglio, "Inside the Shadow War Fought by Russian Mercenaries," BuzzFeed, April 17, 2019, https://www.buzzfeednews.com/article/mikegiglio/inside-wagner-mercenaries-russia-ukraine-syria-prighozhin.

235 Giglio, "Inside the Shadow War Fought by Russian Mercenaries."

236 https://syria360.wordpress.com/2015/10/02/syria-in-the-last-24-hours-russian-and-syrian-airstrikes-crush-daesh/

237 Interview, Amman, March 28, 2019.

238 Jean-Charles Brisard and Kévin Jackson, "The Islamic State's External Operations and the French-Belgian Nexus," *CTC Sentinel* 9, no. 11 (November/December 2016): 8–15, https://ctc.usma.edu/november-december-2016/. From: chrome-extension://oemmndcbldboiebfn-

laddacbdfmadadm/http://cat-int.org/wp-content/uploads/2016/11/
ISIS-Opex-and-the-French–Belgian-Nexus.pdf (Abu Muhammad al-
Adnani, "Say, 'die in your rage,'" pietervanostaeyen.wordpress.com,
January 26, 2015).

239 https://icct.nl/publication/report-the-foreign-fighters-phenomenon-
in-the-eu-profiles-threats-policies/

240 Interview, December 15, 2018.

241 Interview, July 4, 2019.

242 Florence Gaub and Julia Lisiecka, *The Crime-Terrorism Nexus* (Paris:
European Union Institute for Security Studies, April 1, 2017).

243 Raphael Satter and John-Thor Dahlburg, "Paris Attacks: Belgian Abdelhamid Abaaoud Identified as Presumed Mastermind," CBC, November
16, 2015, https://www.cbc.ca/news/world/paris-attacks-mastermind-
abaaoud-1.3320483.

244 https://steamcommunity.com/groups/5eHuss

245 "Attaques à Paris: On pensait que c'étaient des pétards. C'étaient des
scènes de guerre," *Le Monde*, November 14, 2015.

246 Andrew Higgins and Milan Schreuer, "Attackers in Paris 'Did Not Give
Anybody a Chance,'" *New York Times*, November 14, 2015.

247 Eric Reguly, "Two Dead, Eight Arrested after Police Raid Paris Apartment in Hunt for Suspects," *Globe and Mail*, November 17, 2015.

248 Robert-Jan Bartunek and Alastair Macdonald, "Belgians Seize Key
Suspects in Paris, Brussels Attacks," Reuters, April 8, 2016.

249 Alissa J. Rubin, Aurelien Breeden, and Anita Raghavan, "Strikes
Claimed by ISIS Shut Brussels and Shake European Security," *New
York Times*, March 22, 2016.

250 Interview, JAF Special Forces headquarters, Amman, June 13, 2019.

251 Interview, Amman, June 14, 2019.

252 Julie Hirschfeld Davis, "Obama Reports Gains, and 'Momentum,' against ISIS," *New York Times*, April 13, 2016.

253 Michael S. Schmidt and Mark Landler, "U.S. Will Deploy 560 More Troops to Iraq to Help Retake Mosul from ISIS," *New York Times*, July 11, 2016.

254 Interview, Amman, June 14, 2019.

255 Interview, JAF Special Forces headquarters, Amman, June 13, 2019.

256 Interview, JAF Special Forces headquarters, Amman, June 13, 2019.

257 Interview, JAF Special Forces headquarters, Amman, June 13, 2019.

258 Interview, Amman, June 14, 2019.

259 Interview, JAF Special Forces headquarters, Amman, June 13, 2019.

260 Michael Weiss and Pierre Vaux, "Russia's Wagner Mercenaries Have Moved Into Libya. Good Luck With That," The Daily Beast, September 12, 2019.

261 Martin H. Manser, ed., *The Westminster Collection of Christian Quotations* (Louisville: Westminster John Knox Press, 2001), 68.

262 Barbara Starr, "Army's Delta Force Begins to Target ISIS in Iraq," CNN, February 29, 2016.

263 Helene Cooper, Eric Schmitt, and Michael S. Schmidt, "U.S. Captures ISIS Operatives, Ushering in Tricky Phase," *New York Times*, March 1, 2016.

264 Ann Scott Tyson, "Anatomy of the Raid on Hussein's Sons," *Christian Science Monitor*, July 24, 2003.

265 Eric Schmitt and Carolyn Marshall, "In Secret Unit's 'Black Room,' a Grim Portrait of U.S. Abuse," *New York Times*, March 19, 2006.

266 Ian Cobain, "Camp Nama: British Personnel Reveal Horrors of Secret US Base in Baghdad," *Guardian*, April 1, 2013.

267 "McChrystal's Men: TF 6-26," Daily Dish, *Atlantic*, May 15, 2009.

268 chrome-extension://oemmndcbldboiebfnladdacbdfmadadm/
     https://www.inherentresolve.mil/Portals/14/Documents/Strike%20
     Releases/2016/03March/20160304%20Strike%20Release%20Final.
     pdf?ver=2017-01-13-131258-280

269 https://www.inherentresolve.mil/Media-Library/Article/689399/
     pentagon-provides-details-on-target-of-march-4-syria-airstrike/

270 Eric Schmitt and Michael S. Schmidt, "Omar the Chechen, a Senior
     Leader in ISIS, Dies after U.S. Airstrike," *New York Times*, March 15,
     2016.

271 Interview, Amman, GID headquarters, June 14, 2019.

272 Interview, Amman, GID headquarters, June 14, 2019.

273 Interview, Jerash Governorate, June 13, 2019.

274 https://www.youtube.com/watch?v=kuEER8moGe4

275 Interview, General S. Barzani, Makhmur, Kurdistan, June 11, 2019.

276 Hamza Hendawi, Qassim Abdul-Zahra, and Bassem Mroue, "ISIS Has
     'Special Forces,'" *Business Insider*, July 8, 2015, https://www.business-
     insider.com/isis-has-special-forces-2015-7.

277 Interview, Abu H., Amman, GID headquarters, June 14, 2019.

278 Interview, Amman, June 13, 2019.

279 "Pentagon Admits 'Omar the Chechen' Died This Week, Not Earlier,"
     Radio Free Europe / Radio Liberty, July 15, 2016, https://www.rferl.
     org/a/pentagon-admits-omar-the-chechnyan-died-this-week-not-in-
     march-air-strike-mosul-iraq/27859533.html.

280 chrome-extension://oemmndcbldboiebfnladdacbdfmadadm/
     https://www.inherentresolve.mil/Portals/14/Documents/Strike%20
     Releases/2016/07July/20160711%20Strike%20Release%20Final.
     pdf?ver=2017-01-13-131334-623

281 Robin Wright, "Abu Muhammed al-Adnani, the Voice of ISIS, Is
     Dead," *New Yorker*, August 30, 2016.

282 Feras Hanoush, "ISIS Spokesman Abu Mohammad al-Adnani," Atlantic Council, August 26, 2016.

283 Yasmina Allouche, "The 1982 Hama Massacre," Middle East Monitor, February 12, 2018.

284 Interview, Amman, June 18, 2019.

285 Feras Hanoush, "ISIS Spokesman Abu Mohammad al-Adnani," Atlantic Council, August 26, 2016.

286 Feras Hanoush, "Who Was the ISIS Propaganda Chief Killed by a U.S. Air Strike?," *Newsweek*, August 30, 2016.

287 Tara John, "Everything We Know about ISIS Spy Chief Abu Mohammad al-Adnani," *Time*, August 30, 2016.

288 Interview, Amman, June 18, 2019.

289 Maher Chmaytelli, Stephen Kalin, and Ali Abdelaty, "Islamic State Calls for Attacks on the West during Ramadan in Audio Message," Reuters, May 21, 2016.

290 Tara John, "Everything We Know about ISIS Spy Chief Abu Mohammad al-Adnani," *Time*, August 30, 2016.

291 Interview, Amman, June 18, 2019.

292 Samia Nakhoul and Angus McDowall, "Death of Islamic State's Tactician Comes at a Critical Moment," Reuters, August 31, 2016.

293 James Srodes, "Allen Dulles's 73 Rules of Spycraft," *Intelligencer: Journal of U.S. Intelligence Studies* (Fall 2009): 53.

294 Srodes, "Allen Dulles," 53.

295 "IS Announces Death of Spokesman Abu Muhammad al-'Adnani," Jihadist News, Amaq News Agency, August 30, 2016.

296 Interview Amman, June 14, 2019.

297 Samia Nakhoul and Angus McDowall, "Death of Islamic State's Tactician Comes at a Critical Moment," Reuters, August 31, 2016.

298 Eric Schmitt and Anne Barnard, "Senior ISIS Strategist and Spokesman Is Reported Killed in Syria," *New York Times*, August 30, 2016.

299 Joshua Keating, "Everyone Wants Credit for Killing ISIS's Spokesman," *Slate*, September 1, 2016.

300 Ivan Nechepurenko, "Russia Claims Credit for Killing Senior ISIS
Leader in Syria," *New York Times*, August 31, 2016.

301 https://dod.defense.gov/News/News-Releases/News-Release-View/
Article/930843/statement-by-pentagon-press-secretary-peter-cook-
on-precision-airstrike-targeti/

302 Interview, Muwaffaq Salti Air Base, September 16, 2018.

303 Interview, Amman, June 15, 2019.

304 Interview, Amman, September 17, 2018.

305 https://www.youtube.com/watch?v=eWJNjEUzk0A

306 Wladimir van Wilgenburg, "Millionaire Commander Protects Kurdish Capital Against IS," Middle East Eye, January 18, 2016.

307 Erica Goode, "Kurds Mark 20th Anniversary of Deadly Gas Attack,"
New York Times, March 17, 2008.

308 http://heevie.org/aboutkurdistan

309 Interview, Makhmur, June 11, 2019.

310 Interview, Erbil, June 11, 2019.

311 Bruce Campion-Smith, "Canada's Special Forces Play Vital Role in
Helping Peshmerga Troops Battle Daesh in Iraq," *Star*, April 29, 2016.

312 Interview, Makhmur, June 11, 2019.

313 Nick Paton Walsh et. al., "Battle for Mosul Begins with Gunfire and
Car Bombs," CNN, October 17, 2016.

314 Interview, Erbil, June 10, 2019.

315 Interview, Erbil, June 10, 2019.

316 https://www.youtube.com/watch?v=KbsesrAMjTw

317 Interview, Makhmur, June 11, 2019.

318 Interview, Makhmur, June 11, 2019.

319 Interview, Amman, June 17, 2019.

320 "Intelligence Reform on the Cards," Intelligence Online, January 11, 2017.

321 Peter Bergen, "ISIS leaders May Flee Mosul as Their Ranks are Decimated," CNN, October 27, 2016.

322 Interview, Amman, June 14, 2019.

323 Interview, GID headquarters, June 13, 2019.

324 https://www.hbo.com/documentaries/unmasking-jihadi-john-anatomy-of-a-terrorist

325 Interview, Erbil, Peshmerga Ministry, June 10, 2019.

326 https://www.youtube.com/watch?v=5woZG9fQtqo&t=3s

327 "King Appoints New Intelligence Chief, Orders Changes," *Jordan Times*, March 30, 2017.

328 https://kingabdullah.jo/en/interviews/interview-his-majesty-king-abdullah-ii-173

329 Fadi A. Haddadin, "The 'Shia Crescent' and Middle East Geopolitics," *Foreign Policy Blogs* (blog), Foreign Policy Association, January 31, 2017, https://foreignpolicyblogs.com/2017/01/31/shia-crescent-middle-east-geopolitics/.

330 Interview, GID headquarters, June 14, 2019.

331 Mohammed Ghazal, "6 Troops Killed, 14 Injured in Car Bomb Attack on Syria Border," *Jordan Times*, June 21, 2016.

332 Kosar Nawzad, "ISIS Torches Village in Nineveh as MP Warns of Group's Growing Activity," Kurdistan24, April 22, 2019, https://www.kurdistan24.net/en/news/4b8b04fe-50b6-42be-b950-82213672eb6c.

333 Interview, Jerash Governorate, June 13, 2019.

334 Thomas Joscelyn, "CENTCOM: Three Senior Islamic State Foreign Fighters Killed," *FDD's Long War Journal*, May 26, 2017.

335 Interview, Colonel. S., JAF Special Forces headquarters, Amman, June 13, 2019.

336 "Jordan: Four Security Officers Killed after Storming Building," Al Jazeera, August 13, 2018, https://www.aljazeera.com/news/2018/08/jordan-4-security-officers-killed-storming-building-180812080610180.html.

337 Seth J. Frantzman, "Is Iraq's Nineveh Boiling Over?," *Jerusalem Post*, August 11, 2019.

338 Charles Lister, "Trump Says ISIS Is Defeated. Reality Says Otherwise," *Politico*, March 18, 2019.

339 Ellen Ioanes, "A Scathing New Pentagon Report Blames Trump for the Return of ISIS in Syria and Iraq," *Business Insider*, August 8, 2019.

340 Robin Wright, "The Ignominious End of the ISIS Caliphate," *New Yorker*, October 17, 2017.

341 Petra News Agency, "Aqaba Meetings Open for Coordination of Global Efforts to Counter Terrorism, Violent Extremism," *Jordan Times*, April 28, 2019.

INDEX